Y0-BWR-106

PERGAMON INTERNATIONAL LIBRARY
of Science, Technology, Engineering and Social Studies
The 1000-volume original paperback library in aid of education, industrial training and the enjoyment of leisure
Publisher: Robert Maxwell, M.C.

Computer-Oriented Mathematical Physics

INTERNATIONAL SERIES IN
NONLINEAR MATHEMATICS: THEORY, METHODS AND APPLICATIONS
General Editors: V. Lakshmikantham and C. P. Tsokos
VOLUME 3

Other Titles in the Series

Vol. 1 GREENSPAN
Arithmetic Applied Mathematics

Vol. 2 LAKSHMIKANTHAM and LEELA
Nonlinear Differential Equations in Abstract Spaces

Vol. 4 ZACKS
Parametric Statistical Inference

Other Pergamon Titles of Interest

BEECH
Computer Assisted Learning in Science Education

COOPER
Introduction to Dynamic Programming

KOOPMAN
Search and Screening

POLLARD
An Introduction to the Mathematics of Finance (revised edition)

SIMONS
Vector Analysis for Mathematicians, Scientists and Engineers

STREIT and JOHNSON
Mathematical Structures of Quantum Mechanics

Important Research Journals*

Computers and Mathematics with Applications
Reports on Mathematical Physics

*Free specimen copies available on request.

Computer-Oriented Mathematical Physics

by

DONALD GREENSPAN

Department of Mathematics
The University of Texas at Arlington

PERGAMON PRESS

OXFORD · NEW YORK · TORONTO · SYDNEY · PARIS · FRANKFURT

U.K.	Pergamon Press Ltd., Headington Hill Hall, Oxford OX3 0BW, England
U.S.A.	Pergamon Press Inc., Maxwell House, Fairview Park, Elmsford, New York 10523, U.S.A.
CANADA	Pergamon Press Canada Ltd., Suite 104, 150 Consumers Rd., Willowdale, Ontario M2J 1P9, Canada
AUSTRALIA	Pergamon Press (Aust.) Pty. Ltd., P.O. Box 544, Potts Point, N.S.W. 2011, Australia
FRANCE	Pergamon Press SARL, 24 rue des Ecoles, 75240 Paris, Cedex 05, France
FEDERAL REPUBLIC OF GERMANY	Pergamon Press GmbH, 6242 Kronberg-Taunus, Hammerweg 6, Federal Republic of Germany

First edition 1981

British Library Cataloguing in Publication Data

Greenspan, Donald

Computer-oriented mathematical physics.-
(International series in nonlinear mathematics;
v. 3)
1. Mechanics
I. Title
531 QC125.2 80-42043

ISBN 0-08-026471-9 (Hardcover)
ISBN 0-08-026470-0 (Flexicover)

In order to make this volume available as economically and as rapidly as possible the author's typescript has been reproduced in its original form. This method unfortunately has its typographical limitations but it is hoped that they in no way distract the reader.

Printed and bound in Great Britain by
William Clowes (Beccles) Limited, Beccles and London

Preface

This book is intended as an introductory text for beginning university students and advanced high school students. In it we will develop a computer, rather than a continuum, approach to the mechanics of particles. Thus, we will formulate and study new models of classical physical phenomena using only arithmetic. At those points where Newton and Leibniz found it necessary to apply the analytical power of calculus, we shall, instead, apply the computational power of modern digital computers.

Interestingly enough, our definitions of kinetic energy, potential energy, linear momentum and angular momentum will be identical to those of an n-particle system in continuum mechanics, and we will establish the very same laws of conservation and symmetry. In addition, the simplicity of our approach will enable the reader who has only a minimal mathematical background to understand the mechanisms for such complex phenomena as shock wave development and turbulent fluid flow, and to solve dynamical problems resulting from nonlinear interactions. Advanced modelling, however, which requires the interaction of both long and short range forces (Greenspan (1980)), has not been included because of the present book's elementary level of presentation.

The price we pay for mathematical simplicity is that we must do our arithmetic at high speeds. However, the increasing availability of smaller, relatively inexpensive digital computers makes our approach easily implementable.

Since we will emphasize Newtonian mechanics, it should be noted that even in this day of advanced physical theories, the value of our subject has not been diminished that severely. The reason is that both relativistic mechanics and quantum mechanics are not readily applicable to dynamical problems. Relativity, because it precludes simultaneity, does not even allow the existence of the three-body problem. Quantum mechanics, because it requires the inclusion of additional dimensions for each particle under consideration, is unfeasible for all but the simplest of interactions.

From a pedagogical point of view, Chapters 9 and 10 may be omitted from a one semester course if the instructor intends to assign and discuss most of the computer-oriented exercises. Also, the wording of these exercises can be modified easily for adaptation to the use of hand calculators.

Finally, for their help in the preparation of the manuscript, I wish to thank Sandie Jones, who did the programming, Martha Fritz, who did the illustrations, and

Judy Swenson, Gretchen Fitzgerald, and Mary Ann Crain, who did the typing. And fo
permission to quote freely from my monograph DISCRETE MODELS, Addison-Wesley,
Reading, Mass., 1973, I wish to thank the editors of Addison-Wesley.

Donald Greenspan

Contents

CHAPTER 1

Mathematical and Physical Sciences

1.1 INTRODUCTION

Many mathematicians considered Albert Einstein to be a physicist, while many physcists considered him to be a mathematician. The reasons for such contradictory feelings lie in the same individual differences which prevent human beings from agreeing uniformly on almost anything, let alone on such complex, technical questions as to what is a mathematician or what is a physicist. Nevertheless, it will be to our advantage to have some feeling, or intuition, about the nature of our subject, physics, before we begin our study formally. It is the aim of this chapter to develop such intuition and a convenient starting point is to try to understand the often used, rarely understood, statement that "mathematics is an _exact science_".

1.2 MATHEMATICAL SCIENCE

Let us begin by considering some very human problems related to our attempts to communicate with each other by means of _language_. Though we can communicate thought and feeling through other forms, like art and music, still, language is the most universal form used, and so we will concentrate on it. The problems to be discussed are called semantic problems.

If any particular word, like _ship_, were flashed on a screen before a large number of people, it is doubtful that any two persons would form exactly the same image of a ship. Some people might envision ocean liners, others sailing ships, and a few other, perhaps, hydrofoils. The images of ocean liners, for example, might differ in size, number of smoke stacks, color of flags, or location of radar, were they even to have radar. It follows, similarly, that the meaning of _every_ word is so intimately related to a person's individual experiences that probably _no word has exactly the same meaning to any two people_.

But this is not the only problem, because _it does not appear to be possible for anyone to ever find out what a particular word means to anyone else_. Suppose, for example, that man X asks man Y what the word _ship_ means to him and that man Y replies that a _ship_ is a vessel which moves in, on, or under water. Man X, realizing that even a rowboat tied to a pier is moving by virtue of the earth's rotation, asks man Y to clarify his definition of _ship_ by further defining _to move_. Man Y replies that _to move_ is to relocate from one position to another by such processes as walking, running, driving, flying, sailing, and the like. Man X, for exactness, then asks man Y if by sailing he means the process of navigating a ship which has sails, to which man Y replies yes. "Then", replies man X, "I shall never be able to understand you. You have defined ship in terms of move, move in terms of sailing, and sailing in terms of ship, which was the word originally requiring clarification. You have simply talked around a circle."

The circular process in which men X and Y became involved so quickly is, in fact, one in which we can all become entangled if we constantly require definition of words used in definitions. For the total number of words in all existing languages is finite and it would be merely a matter of time to complete a cycle of this linguistic merry-go-round. Moreover, if such difficulties arise so readily with a concrete concept like ship, one can easily imagine the complexities in trying to describe, say, an emotion, like fear or anger. Yet most of us continue to use language with the tacit feeling that we are, in fact, communicating, while som few, like painters, turn to other forms of communication (which, incidentally, hav their own special problems).

Now, in constructing the language of a mathematical science, the scientist examine the two semantic problems described above and agrees that no two people will ever completely understand what any particular word means to the other. With this supposition, however, the problem of definitions resulting in a circular process can be, and is, avoided as follows. Suppose, says the mathematician, the words

a　　in　　path
by　　is　　point
direction　　move　　the
fixed　　out　　trace

are called basic terms and are stated without definition. We all have ideas and feelings about these words, but rather than attempt to make their meanings precise to each other, we shall simply leave them undefined. Now, let us define a line as the path traced out by a point moving in a fixed direction. Note that the word line is defined only in terms of the basic terms. Next, define a plane as the pat traced out by a line moving in a fixed direction. Note that plane is defined in terms only of line and of basic terms. Now suppose that man X asks mathematicia Y what a plane is. Y responds that a plane is the path traced out by a line moving in a fixed direction. Man X, for clarity, asks mathematician Y what he means precisely by a line, to which Y responds that a line is the path traced ou by a point moving in a fixed direction. Man X, seeking further clarity, asks fo the definition of point, to which the mathematician responds, "Point is an undefined basic term", and there the questioning stops.

Thus, every mathematical science begins with basic terms which are undefined and all other concepts are defined by means only of these. Point is an undefined concept of geometry and positive integer is an undefined concept of algebra. No othe subject treats its notions this way.

But let us look a bit further into the nature of mathematical concepts. Consider, for example, the geometric concept called a straight line. With a pencil and ruler, we have all at one time or another drawn a straight line. But, indeed, hav we really ever drawn a straight line? A mathematical line has no width, while the line we draw with pencil and ruler certainly does have some width, even though on might need a special instrument, like a micrometer, to measure the width. As a matter of fact, the width may even vary as the pencil lead is being used up in the drawing process. Indeed, every physical object has some width and it must follow that the mathematical straight line is an idealized form which exists only in the mind, that is, it is an abstraction. In a similar fashion, it can be shown that all mathematical concepts are idealized forms which exist only in the mind, that is, are abstractions.

So, all mathematical concepts are abstractions which either are undefined or have definitions constructed on basic undefined terms.

After having constructed a system of concepts, the mathematician next seeks a body of rules by which to combine and manipulate his concepts. Thus, mathematical sciences now take on the aspects of a game in that rules of play, which must be followed, have to be enumerated. Each mathematical science has its own rules of play, or, what are technically called assumptions or axioms. The axioms of algebra are indeed quite simple. For example, for the numbers 2, 3 and 5, it is assumed that

$$2 + 5 = 5 + 2$$

$$2 \cdot 5 = 5 \cdot 2$$

$$(2+3) + 5 = 2 + (3+5)$$

$$(2 \cdot 3) \cdot 5 = 2 \cdot (3 \cdot 5)$$

$$2 \cdot (3+5) = 2 \cdot 3 + 2 \cdot 5.$$

In complete abstract form, then, if a, b and c are three positive integers, the algebraist assumes that

$$a + b = b + a \qquad \text{(Commutative axiom of addition)}$$

$$a \cdot b = b \cdot a \qquad \text{(Commutative axiom of multiplication)}$$

$$(a+b) + c = a + (b+c) \qquad \text{(Associative axiom of addition)}$$

$$(a \cdot b) \cdot c = a \cdot (b \cdot c) \qquad \text{(Associative axiom of multiplication)}$$

$$a \cdot (b+c) = a \cdot b + a \cdot c \qquad \text{(Distributive axiom).}$$

The question which immediately presents itself is how does one go about selecting axioms? Historically, axioms were supposed to coincide with fundamental physical concepts of truth. But, as the nineteenth century chemists and physicists began to destroy the previous century's physical truths, the choice of mathematical axioms became a relatively free one. And indeed it is a rather simple matter to show that the axioms stated above for numbers can be false when applied to physical quantities. For example, if a represents sulphuric acid and b represents water, while $a + b$ represents adding sulphuric acid to water and $b + a$ represents adding water to sulphuric acid, then $a + b$ is not equal to $b + a$, because $b + a$ results in an explosion whereas $a + b$ does not.

The final difference between mathematical sciences and all other disciplines lies in the reasoning processes allowed in reaching conclusions. There are basically two acceptable types of reasoning in scientific work, inductive reasoning and deductive reasoning. Let us consider each in turn.

Suppose scientist X injects 100 monkeys with virus Y and does not so inject a control group of 100 monkeys. One week later, ninety monkeys in the first group and only five in the second group contract chicken pox. Scientist X, sensing a discovery, repeats the experiment and finds approximately the same statistical results. Further experiments are made in which various environmental factors like heat, light, proximity of cages, and so forth are varied, and in every case X finds that from 85% to 95% of the monkeys receiving virus Y become ill, while only from 3% to 10% of the control group acquire the disease. Scientist X concludes that virus Y is the cause of monkey chicken pox, and the process of reaching his conclusion by experimentation with control is called inductive reasoning. Note that if, after proving his result, X were to inject only one monkey with virus Y, all that he could say would be that the probability is very high that the monkey will become ill. Indeed it is not absolutely necessary that chicken pox would be contracted.

Suppose now that mathematician X writes down a set of assumptions, two of which are

Axiom 1. All heavenly bodies are hollow.

Axiom 2. All moons are heavenly bodies.

Then it *must* follow, *without exception*, that

Conclusion: All moons are hollow.

The above type of reasoning from axioms to *necessary* conclusions is called deductive reasoning. The simple three line argument presented above is called a syllogism. The general process of reaching necessary conclusions from axioms is called deductive reasoning and the syllogism is the fundamental unit in all complex deductive arguments. In mathematics, *all* conclusions must be reached by deductive reasoning alone. Although, very often, *axioms* are selected after extensive inductive reasoning, no mathematical *conclusion* can be so reached. Thus, there is no question of a mathematical conclusion having a high probability of validity as in the case of inductive conclusions. Indeed, if the axioms are absolute truths, then so are the deductive conclusions.

Thus we see that a mathematical science deals with abstract idealized forms which are defined from basic undefined terms, relates its concepts by means of axioms, and establishes conclusions only by deductive reasoning from the axioms. It is the perfect precision of abstract forms and deductive reasoning which qualifies mathematics as being *exact*, while it is the prescribing of materials and methods from which one can create meaningful new forms which gives each mathematical discipline the form of an art.

1.3 PHYSICAL SCIENCE

In speaking of physical science, we will concentrate only on the most basic one, namely, physics. Physics is the study of the fundamental laws of nature. In older times the subject was called *natural philosophy*. The major problem in physics is to discover exactly what the fundamental laws are.

Physicists work in the spirit that anything known is only an approximation and that approximations require constant improvement. The experimental physicist works to improve our knowledge of natural laws by experimenting, deducing, and predicting. The theoretical physicist works in a mathematical spirit to improve our knowledge by dreaming up new models, deducing, and predicting, Theoretical and experimental physicists are in constant interaction in a never ending process of improvement by refinement.

This refinement process has led to the existence today of three major branches of physics, namely, classical physics, relativistic physics, and quantum physics. In classical physics one studies quantities whose gross physical properties are relatively easily observable, and hence have been studied for a long time. These include gases, liquids, solids, sound, light, heat, electricity, magnetism and systems of moving bodies (including the planets). The discovery that the laws of classical physics did not hold when objects moved close to the speed of light led to relativistic physics, in which, for example, one develops a deeper understanding of electrical and magnetic interrelationships. The discovery that the laws of classical physics also did not hold on the atomic and subatomic levels led to quantum physics, in which, for example, one develops a deeper understanding of the nature of the electron and of molecular bonding. Classical physics, the oldest of the three, can be shown to be a limiting case of the other two, which themselves bear little or no relationship to each other. Finally, it is important to note that theoretically all three branches of physics have been formulated as mathematical sciences.

In this book we will concentrate only on classical mechanics. Theoretical discussions will be in the spirit of a mathematical science, since all theoretical physics is mathematical science. Indeed the terms mathematical physics and theoretical physics mean the same thing and can be used interchangeably. However, in order not to obscure the physics, we will not insist on exhibiting the degree of rigor demanded by a mathematician, though this can be done. After all, we are studying physics, not mathematics. This is the way most theoretical physicists work.

1.4 EXERCISES - CHAPTER 1

1.1 Discuss the advantages and disadvantages of human communication by means of painting.

1.2 Discuss the advantages and disadvantages of human communication by means of music.

1.3 Can humans communicate with other classes of animals? If so, with which classes of animals, and how can one test that there is such communication?

1.4 How can one test whether or not humans can communicate with plants?

1.5 To what extent has the mathematical device of beginning with undefined terms really solved the problems inherent in communication by means of language?

1.6 Assuming that addition of two numbers is an undefined operation, show how to define subtraction of two numbers in terms of addition.

1.7 Assuming that multiplication of two numbers is an undefined operation, show how to define division in terms of multiplication.

1.8 Is it possible to add three numbers without adding two of them first and then adding this result to the third number? What then is the significance of the associative law of addition?

1.9 Is it possible to multiply three nonzero numbers without multiplying two of them first and then multiplying the third number by this result? What then is the significance of the associative law of multiplication?

1.10 Euclidean geometry is founded on ten axioms, the most controversial of which can be stated as follows: Through a point not on a given line, one and only one parallel can be drawn to the given line. Why do you think that this axiom has caused such controversy?

1.11 List five conclusions of a scientific nature that were reached by inductive reasoning.

1.12 List five conclusions of a personal nature that you have reached by inductive reasoning.

1.13 Describe a scientific conclusion reached in the past by inductive reasoning which, today, is known to be false.

1.14 Give the necessary conclusions implied by each of the following pairs of axioms, when such a necessary conclusion is implied:

(a) $x = -2y^2 + 4y + 3.$
$y = 1.$

(b) ΔA is similar to ΔB.
ΔB is similar to ΔC.

(c) Men are animals.
Abe is a man.

(d) Each Y is a Z.
Each X is a Y.

(e) Men are animals.
Mean are mortals.

(f) Each X is a Z.
Each X is a Y.

(g) Men are animals.
Mary is a woman.

(h) All red trucks are fire trucks.
All fire trucks are red trucks.

(i) Each X is a Y.
Each Y is an X.

(j) Genes are amino acid structures.
My friend's name is Gene.

(k) The drive to survive dominates animal behavior.
Men and women are animals.

(l) Only communists subvert governments.
The CIA subverted the government of Chile.

(m) Nothing changes.
Everything changes.

1.15 The pleasure of using deductive reasoning to solve puzzles is the basis of much of the interest in "detective" novels. The following, then, are offered for prospective sleuths:

(a) Three men are seated at a round table. From a bag containing three black hats and two white ones, a hat is placed on each man's head. No man can see his own hat, but the other two are completely visible and he does know the original contents of the bag. When asked if he can determine what color hat is on his head, the first man answers "No". Next, when asked if he can determine what color hat is on his head, the second man also answers "No". Finally, when asked if he can determine what color hat is on his head, the third man, who is blind, answers "Yes". What was the color of the third man's hat? Prove your answer.

(b) A man with thirty-one dominoes notices that each domino covers exactly two squares of his chess board. Since a chess board has sixty-four squares, he cuts off diagonally opposite corners so that the cut board has exactly sixty-two squares. Without breaking the dominoes, can he now cover the cut board completely with the dominoes? Prove your answer.

(c) A man has twelve coins, each of which is identical in appearance with the others, but one of which is of a different weight than the others. By using a balance no more than three times, find the bad coin and identify whether it is lighter or heavier than the others.

CHAPTER 2

Gravity

2.1 INTRODUCTION

To begin anything in the "right way" is usually very difficult. Often we begin, make some errors, see what is right, and then begin again. Sometimes we have to make many errors. Sometimes we never see what is right.

We are faced now with exactly such a problem, namely: How shall we model physical behavior in the right way by using only arithmetic? In order to give an answer to this question, we shall explore and analyze a simple experiment related to gravity We choose to study gravity because there is a general awareness of it and because its effects are easy to observe.

2.2 A SIMPLE EXPERIMENT

Things fall. We often say that the reason things fall is gravity. In fact, not everything falls, as was indicated by the first Sputnik. But things "close" to earth do fall, so let us consider only these for the present.

To discover the details of falling, consider the following simple experiment. Let us drop a solid metal ball, which we denote by P, from a height x_0 above ground and measure its height x above ground every Δt seconds as it falls. For example, if we had a camera whose shutter time was Δt, we could first take a sequence of pictures at times $t_0 = 0$, $t_1 = 1(\Delta t)$, $t_2 = 2(\Delta t)$, $t_3 = 3(\Delta t), \ldots,$ or, more concisely, at the distinct times $t_k = k(\Delta t)$, $k = 0,1,2,\ldots,n$.

From the knowledge of the initial height x_0, one could then determine the heights $x_1 = x(t_1)$, $x_2 = x(t_2)$, $x_3 = x(t_3), \ldots,$ or, more concisely, $x_k = x(t_k)$, $k = 0,1,2,\ldots,n$, directly from the photographs by elementary ratio and proportion. For example, suppose $x_0 = 400$ is the height of a building from which the ball has been dropped. As shown in Fig. 2.1, let an X-axis be superimposed such that its origin is at the base of the building. Label the top of the building T and, from the photograph, determine the ratio $|PO| : |TO|$. Then the actual height x of P is determined readily from

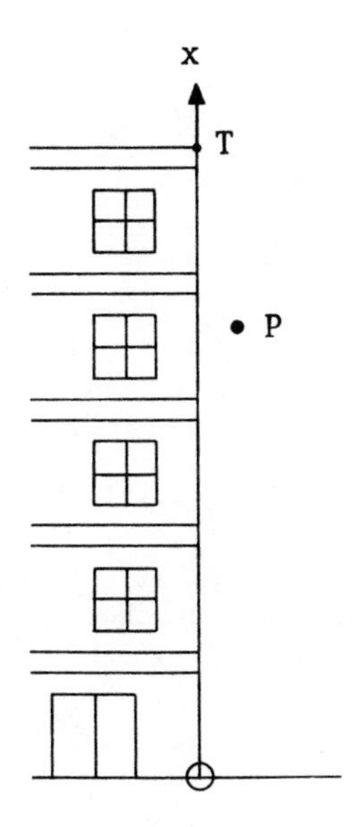

FIGURE 2.1

$$x : 400 = |PO| : |TO|.$$

In this manner, suppose that for $\Delta t = 1$ one finds, to the nearest foot, that

$$
\begin{aligned}
x_0 &= 400 \\
x_1 &= 384 \\
x_2 &= 336 \\
x_3 &= 256 \\
x_4 &= 144.
\end{aligned}
\tag{2.1}
$$

(Note that we will make no special choice of units. One should be able to think in any convenient set of units. Though feet are used in the present discussion, metric units will be used later.) In order to analyze our data, consider first rewriting them in a fashion which exhibits clearly the distance P has fallen. Thus,

$$
\begin{aligned}
x_0 &= 400 = 400 - 0 \\
x_1 &= 384 = 400 - 16 \\
x_2 &= 336 = 400 - 64 \\
x_3 &= 256 = 400 - 144 \\
x_4 &= 144 = 400 - 256.
\end{aligned}
\tag{2.2}
$$

But simple factoring now reveals that

$$
\begin{aligned}
x_0 &= 400 - 16(0)^2 \\
x_1 &= 400 - 16(1)^2 \\
x_2 &= 400 - 16(2)^2 \\
x_3 &= 400 - 16(3)^2 \\
x_4 &= 400 - 16(4)^2,
\end{aligned}
\tag{2.3}
$$

so that, more concisely,

$$x_k = 400 - 16(t_k)^2, \quad k = 0,1,2,3,4. \tag{2.4}$$

It is most important to observe that (2.4) is a formula which has been deduced from the given measurements. We had no idea, at the outset, whether or not any formula existed which related each x_k to each t_k, $k = 0,1,2,3,4$.

One might now be quite eager to assert that (2.4) is a <u>general</u> formula for the position of P at <u>any</u> time, not just at t_0, t_1, t_2, t_3, and t_4. This then requires further experimentation to accumulate more data. For example, from (2.4) one would <u>predict</u> for $k = 5$ that

$$x_5 = 400 - 16(5)^2 = 0,$$

which is, indeed, correct.

2.3 VELOCITY

In order to analyze the motion described in Section 2.2 in greater detail, it might be convenient to discuss next how fast the particle moved. In this connection, we will at present use the terms <u>velocity</u> and <u>speed</u> interchangeably. Later, we shall be more discerning.

Our first problem is to devise some method by which we can determine P's velocities v_0, v_1, v_2, v_3, v_4 at the respective times t_0, t_1, t_2, t_3, t_4, that is, $v_k = v(t_k)$, $k = 0,1,2,3,4$. With no more than a camera available, this cannot be done experimentally. Instead of introducing new equipment into our experiment, let us see if we can devise some useful formula for this purpose.

Since P was dropped from a position of rest, it is reasonable to assume that its initial velocity is zero, that is, $v_0 = 0$. Consider, next, v_1. The velocity v_1 at time t_1 must represent a measure of how fast P's height is changing with respect to time. And, since P's height has changed $x_1 - x_0$ in the time $t_1 - t_0 = \Delta t$, let us define v_1 by the simple formula

$$\frac{v_1+v_0}{2} = \frac{x_1-x_0}{\Delta t}, \tag{2.5}$$

which is equivalent to

$$v_1 = -v_0 + \frac{2}{\Delta t}(x_1-x_0). \tag{2.6}$$

The formula (2.5) is superior to various other possible formulas, like

$$v_1 = \frac{x_1-x_0}{\Delta t}, \tag{2.7}$$

in two very important ways. First, the left side of (2.5) is an averaging, or smoothing, formula, for the velocities, and averaging, or smoothing, is of exceptional value in the analysis of approximate data. Second, formula (2.5) includes v_0. Never mind that v_0 was zero in our experiment, for we certainly did not have to drop P initially, we could actually have thrown it, in which case v_0 would not have been zero. The important point is that, at time t_1, v_0 is known and (2.5) includes this additional bit of knowledge.

Having now defined v_1, it is natural to define $v_2, v_3, v_4, \ldots$, in the following way if we wish to make maximum use of modern digital computer capabilities. The numbers 0 and 1 in (2.5) are called indices and modern computers can increase these indices by unity any number of times with exceptional speed. Thus, to utilize the computer efficiently, it would be desirable to define v_2 by

$$\frac{v_2+v_1}{2} = \frac{x_2-x_1}{\Delta t},$$

to define v_3 by

$$\frac{v_3+v_2}{2} = \frac{x_3-x_2}{\Delta t},$$

and so forth. Thus, in general, we define v_{k+1}, $k = 0,1,2,\ldots,n$, by

$$\frac{v_{k+1}+v_k}{2} = \frac{x_{k+1}-x_k}{\Delta t}, \quad k = 0,1,2,\ldots,n, \tag{2.8}$$

which is, of course, equivalent to

$$v_{k+1} = -v_k + \frac{2}{\Delta t}(x_{k+1}-x_k). \tag{2.9}$$

Data (2.1), $\Delta t = 1$, $v_0 = 0$, and formula (2.9) now yield

$$\begin{aligned}
v_0 &= 0 \text{ ft/sec} \\
v_1 &= -v_0 + 2(x_1-x_0)/(\Delta t) = 0 + 2(384-400)/1 = -32 \text{ ft/sec} \\
v_2 &= -v_1 + 2(x_2-x_1)/(\Delta t) = 32 + 2(336-384)/1 = -64 \text{ ft/sec} \\
v_3 &= -v_2 + 2(x_3-x_2)/(\Delta t) = 64 + 2(256-336)/1 = -96 \text{ ft/sec} \\
v_4 &= -v_3 + 2(x_4-x_3)/(\Delta t) = 96 + 2(144-256)/1 = -128 \text{ ft/sec}
\end{aligned} \tag{2.10}$$

2.4 ACCELERATION

To analyze the motion in our experiment in still greater detail, it may be of value to examine how fast the velocity is changing, or what is called the acceleration. If P's acceleration is a_0, a_1, a_2, a_3 and a_4 at the respective times t_0, t_1, t_2, t_3, t_4, then it would seem natural, in the spirit of Section 2.3, to use a general formula analogous to (2.8), namely,

$$\frac{a_{k+1}+a_k}{2} = \frac{v_{k+1}-v_k}{\Delta t}, \tag{2.11}$$

or, equivalently,

$$a_{k+1} = -a_k + \frac{2}{\Delta t}(v_{k+1}-v_k). \tag{2.12}$$

Recursive formula (2.12) will yield a_1, a_2, a_3, and a_4 readily, provided we know a_0. But with some thought it becomes clear that we do not know a_0. We knew x_0 because it was given, and we decided that it was reasonable to assume $v_0 = 0$ because P was dropped from a position of rest. There simply doesn't seem to be any facet of the experiment which enables us to make a reasonable choice for a_0. So, we must discard (2.11) and (2.12) and rethink the matter.

Since acceleration is to be a measure of how the velocity is changing with respect to time and since a_0 is not known, let us define a_0 by the very simple formula

$$a_0 = \frac{v_1-v_0}{\Delta t}.$$

But now that a_0 has been defined, it is computationally most convenient to define a_1, a_2, a_3 and so forth by the unitary increment of indices, that is, by

$$a_1 = \frac{v_2-v_1}{\Delta t}$$

$$a_2 = \frac{v_3-v_2}{\Delta t}$$

$$a_3 = \frac{v_4-v_3}{\Delta t}.$$

Thus, in general, we define a_k, $k = 0,1,2,\ldots,n$, by

$$a_k = \frac{v_{k+1}-v_k}{\Delta t}, \quad k = 0,1,2,\ldots,n. \tag{2.13}$$

For our experiment of Section 2.2, we have from (2.10) and (2.13)

$$a_0 = (v_1-v_0)/\Delta t = (-32+0)/1 = -32 \text{ ft/sec}^2$$

$$a_1 = (v_2-v_1)/\Delta t = (-64+32)/1 = -32 \text{ ft/sec}^2$$

$$a_2 = (v_3 - v_2)/\Delta t = (-96+64)/1 = -32 \text{ ft/sec}^2$$

$$a_3 = (v_4 - v_3)/\Delta t = (-128+96)/1 = -32 \text{ ft/sec}^2. \quad (2.14)$$

Thus, (2.14) yields the remarkable result that, under the influence of gravity, P's acceleration at t_0, t_1, t_2, t_3 is constant, with the exact value -32 ft/sec^2.

2.5 FURTHER EXPERIMENTS

The deductions reached thus far suggest interesting, new questions which require additional experimentation and analysis. Many of these questions deal with the effect on the motion of P when various initial conditions change. Thus, for example, repetition of the experiment with five metal balls of different weights, dropped from several different initial heights yields, in each case, exactly the same conclusions as in Sections 2.2 - 2.4. Use of a high speed camera to decrease the time step Δt also yields nothing new. But, when a wooden ball is dropped, the results are slightly different, and, when a light paper ball and a paper plate are dropped, the results are distinctly different. So, weight and shape seem to make a difference, and the question is why. At this point, it doen't take very much imagination to realize that the presence of air does have some effect. This is especially noticeable from the erratic motion of the paper plate as it falls. So, we experiment once again, but this time in a laboratory, where we can pump the air out of a long vertical tube, thus creating a partial vacuum. As each of the objects listed above falls in this tube, we see that all the conclusions of Section 2.2 - 2.4 are valid. Indeed, in this tube even a small feather falls in exactly the same way as a metal ball.

2.6 A MATHEMATICAL MODEL

Let us see now if we can develop a theory for gravity. By this we will mean the following. Assuming only the barest essentials, let us try to prove all the other results found thus far, and, perhaps, even more. The "even more" results will then serve as theoretical predictions. Experimental verification of these predictions will add credence to the theory, while disproof of the predictions will require a theoretical reformulation. The intuition used in the discussion which follows is derived from the development in Sections 2.2 - 2.5.

Let a particle P at height x_0 be dropped in a vacuum from a position of rest. For $\Delta t > 0$, let $t_k = k\Delta t$, $k = 0,1,2,\ldots,n$. At each time t_k, let the particle's position, velocity, and acceleration be x_k, v_k and a_k, respectively. Assume that $v_0 = 0$ and that

$$\frac{v_{k+1}+v_k}{2} = \frac{x_{k+1}-x_k}{\Delta t}, \quad k = 0,1,2,\ldots,n \quad (2.15)$$

$$a_k = \frac{v_{k+1}-v_k}{\Delta t}, \quad k = 0,1,2,\ldots,n. \quad (2.16)$$

Finally, assume that

$$a_k = -32 \text{ ft/sec}^2, \quad k = 0,1,2,\ldots,n. \tag{2.17}$$

We wish now to prove that the motion of P is exactly as was discovered in Sections 2.2 - 2.5. We will do this by showing how the above assumption determine completely the motion of P from the fixed initial position x_0. To proceed, how ever, will require some observations and results in the use of summation symbols.

The symbol $\sum_{k=i}^{n} f(x_k)$, read "the summation from k equals i to k equals n of $f(x_k)$" means, let $k = i, i+1, i+2, \ldots, n$ in $f(x_k)$ and then add up all the resulting f's. Thus,

$$\sum_{k=1}^{5} x_k = x_1 + x_2 + x_3 + x_4 + x_5$$

$$\sum_{k=0}^{5} [(-1)^k y_k] = y_0 - y_1 + y_2 - y_3 + y_4 - y_5$$

$$\sum_{k=2}^{7} (k)^2 = 2^2 + 3^2 + 4^2 + 5^2 + 6^2 + 7^2.$$

The k in $\sum_{k=i}^{n} f(x_k)$ is called a "dummy index", since it can be changed to any other (unused) letter without changing the actual sum. For example,

$$\sum_{k=1}^{5} x_k^2 = x_1^2 + x_2^2 + x_3^2 + x_4^2 + x_5^2$$

is identical with

$$\sum_{j=1}^{5} x_j^2 = x_1^2 + x_2^2 + x_3^2 + x_4^2 + x_5^2$$

To facilitate algebraic manipulation with summations, the following identities wil be of exceptional value:

$$\sum_{k=i}^{n} [\alpha f_k + \beta g_k] = \alpha \sum_{k=i}^{n} f_k + \beta \sum_{k=i}^{n} g_k; \quad \alpha, \beta \text{ constants} \tag{2.18}$$

$$\sum_{k=0}^{n-1} (x_{k+1} - x_k) = x_n - x_0 \tag{2.19}$$

$$\sum_{k=1}^{n} 1 = n. \tag{2.20}$$

The validity of (2.18) can be seen clearly from the following particular example:

$$\sum_{k=1}^{4}(3x_k-4Y_k) = (3x_1-4Y_1) + (3x_2-4Y_2) + (3x_3-4Y_3) + (3x_4-4Y_4)$$
$$= 3(x_1+x_2+x_3+x_4) - 4(Y_1+Y_2+Y_3+Y_4)$$
$$= 3\sum_{k=1}^{4} x_k - 4\sum_{k=1}^{4} Y_k.$$

The validity of (2.19) can be seen clearly from the following particular example:

$$\sum_{k=0}^{4}(x_{k+1}-x_k) = (x_1-x_0) + (x_2-x_1) + (x_3-x_2) + (x_4-x_3) + (x_5-x_4)$$
$$= -x_0 + (x_1-x_1) + (x_2-x_2) + (x_3-x_3) + (x_4-x_4) + x_5$$
$$= x_5 - x_0.$$

The cancellation of all terms between x_5 and x_0, above, is called <u>telescoping</u> and (2.19) is called a telescopic sum. The validity of (2.20) follows readily, since

$$\sum_{k=1}^{n} 1 = \underbrace{1 + 1 + 1 + \ldots + 1}_{n \text{ terms}} = n.$$

Let us return now to (2.15) - (2.17). From (2.16) and (2.17) it follows that

$$v_{k+1} - v_k = -32\Delta t.$$

Summing both sides of this equation yields, for each integer $n \geq 1$,

$$\sum_{k=0}^{n-1}(v_{k+1}-v_k) = \sum_{k=0}^{n-1}(-32\Delta t), \quad n = 1,2,3,\ldots,$$

or, by (2.18) - (2.20),

$$v_n - v_0 = -32\, n\Delta t.$$

Thus, since $v_0 = 0$ and $n\Delta t = t_n$, it follows that

$$v_n = -32t_n, \quad n = 1,2,3,4,\ldots \tag{2.21}$$

is the formula for the velocity of P at the time steps $t_1, t_2, t_3, \ldots$. For $\Delta t = 1$, in which case $t_0 = 0$, $t_1 = 1$, $t_2 = 2$, $t_3 = 3$ and $t_4 = 4$, formula (2.21) yields the exact values determined in (2.10). Observe also that these values have been deduced without any knowledge or use of x_0. That is, formula (2.21) is valid no matter what the initial height x_0 may be.

From (2.21), then,

$$\frac{v_{k+1}+v_k}{2} = \frac{-32t_{k+1}-32t_k}{2} = -16(t_{k+1}+t_k)$$

$$= -16\Delta t[(k+1) + k] = -16\Delta t(2k+1),$$

so that (2.15) can be rewritten as

$$\frac{x_{k+1}-x_k}{\Delta t} = -16\Delta t(2k+1),$$

or, equivalently, as

$$x_{k+1} - x_k = -16(\Delta t)^2(2k+1)$$

Summation of both sides of the latter equation then yields

$$x_n - x_0 = -16(\Delta t)^2 \sum_{k=0}^{n-1} (2k+1), \quad n = 1,2,3,4,\ldots . \tag{2.22}$$

But $\sum_{k=0}^{n-1}(2k+1)$ is the sum of an arithmetic sequence of n terms whose first term is 1 and whose last term is $2n - 1$. This sum is then equal to n^2. Thus (2.22 simplifies to

$$x_n = x_0 - 16(n\Delta t)^2$$

or

$$x_n = x_0 - 16(t_n)^2, \quad n = 1,2,3,4,\ldots, \tag{2.23}$$

which is the formula for the position x_n of P at each time step t_n. For $x_0 = 400$ and $\Delta t = 1$, this formula yields exactly the same positions as those of (2.4).

Now, since we are theorizing, let us see if we can create some new concepts which might be applicable to our analysis of gravity. We know that if there is a large, heavy box sitting in the middle of our living room, we would have to apply some force to move it. Physically speaking, the application of the force would change the box's velocity from zero to a nonzero value. But, a change in velocity is called acceleration. So, if we apply a force, the box will accelerate. Thus, force and acceleration seem to be directly related. Moreover, the heavier the box the more force is needed to move it, so the force is also related directly to the weight, or mass (speaking loosely for the moment) of the box. We wish then to make a guess at the relationship between force F, mass m, and acceleration a, and one such simple possibility, which we will continue to explore theoretically, is

$$F = ma.$$

More precisely, at the time t_k, if F_k and a_k are the related force and acceleration, while the mass m is a constant, we shall take our equation in the form

$$F_k = ma_k, \quad k = 0,1,2,3,\ldots \,. \tag{2.24}$$

Let us now consider the impact of the choice (2.24). If we define a mathematical quantity W_n, $n = 1,2,3,\ldots$, (which, at present, has no physical role, whatsoever, but will be discussed in Section 3.3 and will there be called the work done by the force) by

$$W_n = \sum_{k=0}^{n-1} (x_{k+1}-x_k)F_k, \tag{2.25}$$

then, (2.15), (2.23) - (2.25) yield

$$\begin{aligned} W_n &= \sum_{k=0}^{n-1} (x_{k+1}-x_k)ma_k = m\sum_{k=0}^{n-1} (x_{k+1}-x_k)\left(\frac{v_{k+1}-v_k}{\Delta t}\right) \\ &= m\sum_{k=0}^{n-1}\left(\frac{x_{k+1}-x_k}{\Delta t}\right)(v_{k+1}-v_k) = m\sum_{k=0}^{n-1}\left(\frac{v_{k+1}+v_k}{2}\right)(v_{k+1}-v_k) \\ &= \frac{m}{2}\sum_{k=0}^{n-1}(v_{k+1}^2-v_k^2) = \frac{m}{2}(v_n^2-v_0^2). \end{aligned}$$

Thus,

$$W_n = \frac{m}{2}v_n^2 - \frac{m}{2}v_0^2, \quad n = 1,2,\ldots \,. \tag{2.26}$$

If one defines K_k, the kinetic energy at time t_k, by

$$K_k = \frac{1}{2}mv_k^2, \tag{2.27}$$

then (2.26) and (2.27) imply

$$W_n = K_n - K_0. \tag{2.28}$$

Intuitively speaking, the kinetic energy K_k is a measure of a particle's ability to do work because of its velocity.

Now, for gravity, $a_k = -32$. Thus $F_k \equiv -32m$, so that reconsideration of (2.25) yields now

$$W_n = \sum_{k=0}^{n-1}(x_{k+1}-x_k)(-32m) = -32m\sum_{k=0}^{n-1}(x_{k+1}-x_k) = -32mx_n + 32mx_0$$

If one defines the potential energy V_k at time t_k by

$$V_k = 32mx_k, \tag{2.29}$$

then

$$W_n = -V_n + V_0, \quad n = 1,2,\ldots \,. \tag{2.30}$$

The potential energy, intuitively, is a measure of the particle's ability to do work because of its height above ground.

Finally, elimination of W_n from (2.28) and (2.30) yields

$$K_n - K_0 = -V_n + V_0, \quad n = 1,2,\ldots$$

or

$$K_n + V_n = K_0 + V_0, \quad n = 1,2,\ldots \ . \tag{2.31}$$

The result (2.31) is very interesting because it says that for each value of n, $K_0 + V_0 = K_1 + V_1 = K_2 + V_2 = \ldots$, that is, the sum of the kinetic and potential energies at each time is always the same, namely, $K_0 + V_0$. Thus, as a particle falls, its potential energy changes and its kinetic energy changes, but the sum must always be the same. Thus, a decrease in kinetic energy must yield an increase in potential energy, and vice versa. The result (2.31) is given a special name, the Law of Conservation of Energy.

2.7 STILL FURTHER EXPERIMENTS

The model described in Section 2.7 yielded new, interesting results which were not available previously, but, as usual, raised new questions which require additional experimental studies.

First, one could test easily, and indeed verify, that the conclusion following (2.20) is valid, that is, that the velocity v_n at time t_n of particle P, dropped from rest in a vacuum, is independent of the initial height x_0.

Next, for example, one could question the validity of (2.24). Experimentation does support its use in the study of many motions which we have all seen or experienced, like that of a bicycle, or a pendulum, or a gyroscope, or a large variety of every day machines. To a limited degree, it is useful in the study of planetary motions. But, for motions on a scale larger than the solar system, for motions on a scale smaller than the molecule, and for the study of electromagnetic radiation, it seems to fail rather badly.

One could question also our thinking about gravity. We spoke of things falling to the earth as if the earth were exerting a pull. But the observation that $a = -32$ ft/sec^2 was the same for all falling particles implies, from (2.24), that F/m is constant. Hence, since a change of particles results in a change of m, it must also result in a change in F. But if the earth is exerting this force we call gravity, why should it vary with the particle on which it is being applied. After all, if I decide to push with all my strength, the force I exert is only a matter of how strong I am and does not depend on the object which I may decide to push. So, if (2.24) is valid, thus implying that F varies with the mass of the falling body, something seems to have escaped our notice. Further, it is peculiar that not all objects, like Sputnik and the moon, fall to the earth. These questions will be explored later when we study gravity as a particular instance of gravitation.

We remark finally that attention in this book is confined to motions for which (2.24) is valid. Equation (2.24) is called Newton's equation after its creator,

Isaac Newton (1642-1727). To solve actual problems which result from the application of (2.24), Newton developed the sophisticated mathematical subject called <u>calculus</u>, which, incidentally was developed independently by Gottfried Wilhelm Leibniz (1646-1716). Today, however, with the availability of modern digital computers, we can apply (2.24) using only arithmetic, and it is in this new spirit that we proceed.

2.8 EXERCISES - CHAPTER 2

2.1 Restate (2.1) in metric units. Restate (2.4) in metric units. Is it possible to deduce the metric form of (2.4) from the metric form of (2.1)?

2.2 Write out each of the following sums.

(a) $\sum_{k=0}^{5} y_k$

(b) $\sum_{k=1}^{6} 2y_k$

(c) $\sum_{k=-1}^{5} (y_k^2)$

(d) $\sum_{k=2}^{6} (-1)^k y_k^2$

(e) $\sum_{k=-2}^{2} ky_k$

(f) $\sum_{k=0}^{10} k^2 \sin x_k$

(g) $\sum_{k=-1}^{4} (k \sin y - 2 \cos x_k)$

(h) $\sum_{k=1}^{10} 2\pi$

2.3 Use (2.18) to rewrite each of the following in an equivalent form.

(a) $\sum_{k=1}^{6} [7x_k - 9y_k]$

(b) $\sum_{k=1}^{15} [7x_k^2 - 9e^k]$

(c) $\sum_{k=0}^{20} 5 \sin x_k$

(d) $\sum_{k=1}^{30} (-2k+36k^2)$

(e) $\sum_{k=1}^{30} [(-1)^k + 3k^3]$

(f) $\sum_{k=-1}^{10} [3f_k(x) - 9g(k)]$

(g) $\sum_{k=0}^{10} [\pi \sin x_k + x_k \sin \pi]$

2.4 Show that each of the following is telescopic and, thereby, simplify the sum.

(a) $\sum_{k=0}^{7} (y_{k+1}-y_k)$

(b) $\sum_{k=-1}^{10} (y_{k+1}^2-y_k^2)$

(c) $\sum_{k=2}^{25} (y_{k+1}^3-y_k^3)$

(d) $\sum_{k=0}^{10} [(k+1)^2 - k^2]$

(e) $\sum_{k=1}^{10} [\sin x_{k+1} - \sin x_k]$

(f) $\sum_{k=7}^{1000} [(k-6)^2 - (k-7)^2]$

2.5 Evaluate each of the following sums.

(a) $\sum_{k=1}^{5} 1$ (d) $\sum_{k=1}^{5} 5$

(b) $\sum_{k=0}^{5} 1$ (e) $\sum_{j=1}^{9} (-2)$

(c) $\sum_{k=-5}^{5} 1$ (f) $\sum_{k=0}^{10} \pi$

2.6 Show that if a particle is under the influence of gravity, and if $v_0 \neq 0$, then (2.15) - (2.17) imply

$$v_n = v_0 - 32t_n$$

$$x_n = x_0 + v_0 t_n - 16(t_n)^2$$

2.7 A particle is thrown upward from the ground at 64 ft/sec. Using the results of Exercise 2.6, find how high the particle rises and how long it is in the air.

2.8 A particle is thrown upward from the ground at 640 ft/sec. Using the results of Exercise 2.6, find how high it rises and how long it is in the air.

2.9 A girl on the ground wishes to throw a particle exactly 144 feet into the air. With what initial velocity must she throw it?

2.10 Using the heights x_k given by (2.2) and the velocities v_k given by (2.10), verify the Law of Conservation of Energy (2.31) for each of $k = 1,2,3,4$ when the mass is unity.

CHAPTER 3

Theoretical Physics as a Mathematical Science

3.1 INTRODUCTION

In this chapter we will discuss the theoretical foundations of classical physics as a mathematical science. For the present, attention will be restricted to the motion of a single particle in a single direction, thus formalizing the heuristic ideas developed in our study of gravity. Of course, eventually, we must study motion in three dimensions, and the present discussion will facilitate that study.

The methodology to be developed will be completely arithmetic in nature and will be applied to realistic physical problems, for which crude answers often will be obtainable by means only of pencil and paper. Highly accurate answers and analyses will be attainable only with the aid of a modern digital computer. Throughout, the following characteristics of digital computers will guide our formulation of models and their dynamical equations. Modern computers (a) execute arithmetic operations, store numbers, and retrieve numbers with exceptional speed; (b) store only rational numbers with a finite number of decimal places and execute only approximate arithmetic with such numbers, and (c) are restricted to having a smallest positive number, a largest positive number, and a smallest negative number.

3.2 BASIC MATHEMATICAL CONCEPTS

Not all the arithmetic and algebra which one usually learns will be needed. Moreover, the ways in which these mathematical subjects are usually organized are not quite the ways which we will find most convenient. Hence, let us first reorganize our thinking about elementary mathematics. For precision this will be done now in a formalistic way.

Definition 3.1. For $\Delta x > 0$ and for a fixed constant a, the finite set

$$x_k = a + k\Delta x, \quad k = 0,1,2,\ldots,n$$

is called an R_{n+1} set.

Example 1. For $\Delta x = 0.25$, $a = 0$, and $n = 4$, the set

$$x_k = 0 + k(0.25), \quad k = 0,1,2,3,4$$

is an R_5 set. Specifically, this set consists of the points $x_0 = 0$, $x_1 = 0.25$ $x_2 = 0.50$, $x_3 = 0.75$, and $x_4 = 1.00$. It is, geometrically, the set of points which divides the interval $0 \leq x \leq 1$ of an X-axis into four equal parts. Note that it takes five division points to subdivide an interval into four equal parts.

Example 2. For $\Delta x = 10^{-6}$, $a = 0$, and $n = 10^6$, the set $x_k = k(10)^{-6}$, $k = 0, 1,2,\ldots,10^6$, is an R_{10^6+1} set. Without the aid of special lenses, the plot of this set appears to the naked eye to be no different than the entire real interval $0 \leq x \leq 1$. This "packing" of a large, but finite, number of points to give the same visual effect as continuity, coupled with the ability to manipulate constructively with such sets by means of computers, makes R_{n+1} sets both mathematically and physically attractive.

Definition 3.2. Let $x_k = a + k\Delta x$, $k = 0,1,\ldots,n$, be an R_{n+1} set. If to each x_k in R_{n+1} there corresponds by some rule f a unique value y_k, then f is said to be a discrete function on R_{n+1}.

Example 1. On the R_{10^6+1} set $x_k = k(10)^{-6}$, $k = 0,1,2,\ldots,10^6$, the relationship $y_k = x_k^2$, $k = 0,1,2,\ldots,10^6$, defines a discrete function.

Example 2. On the R_6 set $x_0 = 0$, $x_1 = 0.25$, $x_2 = 0.50$, $x_3 = 0.75$, $x_4 = 1.00$ a discrete function is defined by the listing $y_0 = 1$, $y_1 = -1$, $y_2 = 0$, $y_3 = 7$, $y_4 = 3$.

Discrete functions can be defined by explicit relationships, as given in Example 1 above, or by a complete listing, as given in Example 2, above. Moreover, discrete functions are ideally suited for the mathematical representation of experimental data. To do this, one must realize only that the choice of the letters x, y, and f in Definition 3.2 is only a mathematical convenience. Discrete functions could be defined by $y_k = y(x_k)$, $Z_k = f(x_k)$, $w_k = g(x_k)$, $w_k = F(t_k)$, $x_k = G(t_k)$ and so forth. Thus, the data set (2.1) could be represented mathematically by beginning with the R_5 set

$$t_0 = 0, \quad t_1 = 1, \quad t_2 = 2, \quad t_3 = 3, \quad t_4 = 4$$

and defining on it a discrete function $x_k = G(t_k)$ by the listing

$$x_0 = 400, \quad x_1 = 384, \quad x_2 = 336, \quad x_3 = 256, \quad x_4 = 144.$$

The term "function" will mean "discrete function" throughout, unless indicated otherwise.

Definition 3.3. Let f be a function defined on a given R_{n+1} set. Then the graph of f is the set of number couples (x_k,y_k), $k = 0,1,2,\ldots,n$.

Example. On the R_{10^6+1} set $x_k = k(10)^{-6}$, $k = 0,1,2,\ldots,10^6$, the graph of $y_k = x_k^2$ is given in Figure 3.1. Again, note that the process of packing a

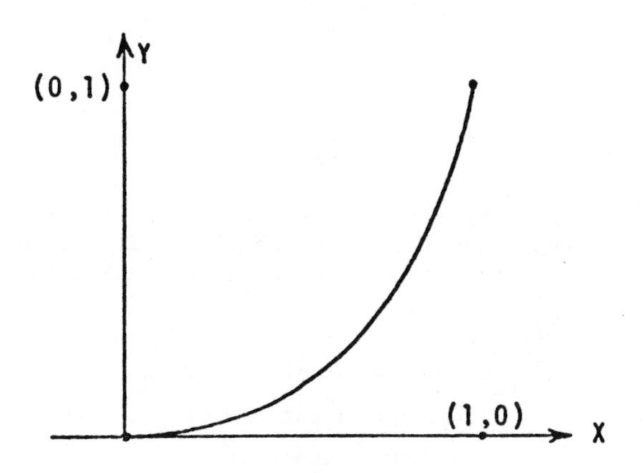

FIGURE 3.1

elatively large, but finite, number of points has resulted in a graph which, to he naked eye, does not differ from that of $y = x^2$ on $0 \le x \le 1$.

ith the basic concepts of set and function now defined, we turn next to the conept of a difference equation. An expression like $y_{k+1} - y_k$ is called a differnce in y's. An expression like $x_{k+1} - x_k$ is called a difference in x's. An xpression like

$$\frac{y_{k+1}-y_k}{x_{k+1}-x_k}$$

s, for obvious reasons, called a difference quotient. Using other symbols, we ave already encountered difference quotients in, for example, (2.8) and (2.13). ote also that if an interval $a \le x \le b$ has been divided into n equal parts by he R_{n+1} set $x_0, x_1, x_2, \ldots, x_n$, then we always have $\Delta x = x_{k+1} - x_k$, $k = 0,1,2,$ $\ldots, n-1$. With this simplification, one can rewrite the above difference quotient s

$$\frac{y_{k+1}-y_k}{\Delta x}.$$

e shall, then, be interested, in particular, in studying equations of the form

$$\frac{y_{k+1}-y_k}{\Delta x} = f(x_k, y_k, y_{k+1}), \quad k = 0,1,2,\ldots,n-1, \tag{3.1}$$

hich are called difference equations.

nfortunately, from the computer point of view, the form of (3.1) is not very conenient. However, if (3.1) is rewritten as

$$y_{k+1} = y_k + (\Delta x) f(x_k, y_k, y_{k+1}), \quad k = 0,1,2,\ldots,n-1,$$

r, equivalently, as

$$y_{k+1} = F(x_k, y_k, y_{k+1}), \quad k = 0,1,2,\ldots,n-1 \tag{3.2}$$

where

$$F(x_k, y_k, y_{k+1}) \equiv y_k + (\Delta x) f(x_k, y_k, y_{k+1}),$$

then the form (3.2) is exceptionally convenient, for (3.2) is what is known as a recursion formula and such formulas are especially suitable for high speed computation. Our definition is then formulated precisely as follows.

Definition 3.4. On an R_{n+1} set $x_0, x_1, \ldots, x_n$, an equation of the form (3.2), where F must depend on y_k, but need not depend on x_k or y_{k+1}, is called a difference equation. If F does not depend on y_{k+1}, then (3.2) is said to be explicit. Otherwise, it is said to be implicit.

As specific examples, on the R_5 set $x_0 = 0$, $x_1 = \frac{1}{4}$, $x_2 = \frac{1}{2}$, $x_3 = \frac{3}{4}$, $x_4 = 1$, the following five equations are difference equations, the first three of which are explicit and the last two of which are implicit:

$$y_{k+1} = y_k^2, \quad k = 0,1,2,3$$

$$y_{k+1} = 2x_k - 3y_k, \quad k = 0,1,2,3$$

$$y_{k+1} = \sin(x_k + y_k), \quad k = 0,1,2,3$$

$$y_{k+1} = \sin(x_k + y_{k+1}) - 3y_k, \quad k = 0,1,2,3$$

$$y_{k+1} = \frac{x_k y_k - 4}{x_k^2 y_k^2 + 1} - 3y_{k+1} + 7, \quad k = 0,1,2,3.$$

Note, incidentally, that the last equation above can be changed into an explicit form by solving for y_{k+1}, but that the next to the last equation cannot be so remanipulated.

With difference equations now defined, it is natural to discuss solutions of such equations.

Definition 3.5. A solution of (3.2) on a given R_{n+1} set is any function $y_k = y(x_k)$, $k = 0,1,2,\ldots,n$, which satisfies (3.2) for $k = 0,1,2,\ldots,n-1$.

It is of essence to understand that we are taking only discrete functions to be solutions of difference equations. Interestingly enough, this natural approach had not been taken before the development of computers.

Example. On the R_5 set $x_0 = 0$, $x_1 = 0.25$, $x_2 = 0.50$, $x_3 = 0.75$, $x_4 = 1.00$ consider the explicit difference equation

$$y_{k+1} = y_k^2 - 4x_k, \quad k = 0,1,2,3. \tag{3.3}$$

One solution of this equation is the discrete function

$$y_0 = 0, \quad y_1 = 0, \quad y_2 = -1, \quad y_3 = -1, \quad y_4 = -2,$$

since

$$y_1 = y_0^2 - 4x_0, \quad \text{or,} \quad 0 = 0^2 - 4(0)$$

$$y_2 = y_1^2 - 4x_1, \quad \text{or,} \quad -1 = 0^2 - 4(0.25)$$

$$y_3 = y_2^2 - 4x_2, \quad \text{or,} \quad -1 = (-1)^2 - 4(0.50)$$

$$y_4 = y_3^2 - 4x_3, \quad \text{or,} \quad -2 = (-1)^2 - 4(0.75).$$

Another solution is

$$y_0 = -1, \quad y_1 = 1, \quad y_2 = 0, \quad y_3 = -2, \quad y_4 = 1.$$

Indeed, an entire family of solutions can be generated from

$$y_0 = C, \quad y_1 = C^2, \quad y_2 = C^4 - 1, \quad y_3 = (C^4-1)^2 - 2,$$

$$y_4 = [(C^4-1)^2 - 2]^2 - 3,$$

by giving C <u>any</u> constant value. The choices $C = 0$ and $C = -1$ yield the first two solutions described above.

As is indicated by the above example, difference equations can have a very large number of solutions. For this reason, we consider next imposing an <u>additional</u> constraint, which can be satisfied by one and only one of the solutions of a given difference equation. For this purpose, we define next an initial value problem.

<u>Definition 3.6</u>. An initial value problem (I.V. problem) for (3.2) is one in which one must find a solution of (3.2) when y_0 is given.

<u>Example</u>. On the R_5 set $x_0 = 0$, $x_1 = 0.25$, $x_2 = 0.50$, $x_3 = 0.75$, $x_4 = 1.00$, the problem of finding a solution of (3.3) for which

$$y_0 = 0 \tag{3.4}$$

is an I.V. problem. Let us show how simple it is, in fact, to generate the solution of this particular problem. For $k = 0$, (3.3) has the particular form

$$y_1 = y_0^2 - 4x_0$$

which, from the given data, yields

$$y_1 = 0.$$

For $k = 1$, (3.3) yields

$$y_2 = y_1^2 - 4x_1$$

which, from the given data and the value of y_1 just determined, yields

$$y_2 = -1.$$

For $k = 2$, (3.3) yields

$$y_3 = y_2^2 - 4x_2,$$

which implies

$$y_3 = -1.$$

Finally, for $k = 3$, (3.3) yields

$$y_4 = y_3^2 - 4x_3,$$

which implies

$$y_4 = -2.$$

Thus, the discrete function

$$y_0 = 0, \quad y_1 = 0, \quad y_2 = -1, \quad y_3 = -1, \quad y_4 = -2$$

is the solution of the given problem. If R_{n+1} were to consist of 10^6 points, we would have employed a computer to generate the solution.

The above discussion results in the following theorem, which assures, a priori, that certain problems have solutions and that these solutions are unique. Before using a computer, this knowledge is valuable to have, since it provides some assurance that we are not wasting our time on a problem for which there may be no solution.

Theorem 3.1. If (3.2) is explicit, then an I.V. problem for (3.2) has a unique solution provided only that F is defined at each stage of iteration (3.2). The solution can be generated on a computer provided that F is a computer function and that no y_{k+1} is in absolute value larger than the largest number in the computer.

If (3.2) is implicit, then an existence and uniqueness theorem like the above is rarely easy to establish. Such matters will be discussed with each specific implicit equation, when it is introduced.

Also, there is in Theorem 3.1 the not-too-subtle hint that, while computing, one's numbers may become larger, in absolute value, than the computer can handle. When this occurs, the computer terminology for it is overflow, while the mathematical terminology is instability. We shall discuss such problems only as they arise.

We have now all the basic mathematics needed for the study of physical problems. Indeed, the amount required is quite minimal. Such simplicity results only because of the availability of computers.

3.3 BASIC PHYSICAL CONCEPTS

As indicated in Chapter I, each mathematical science must begin with basic undefined terms, and since theoretical physics is a mathematical science, we now list our undefined terms to be: particle, mass, force, distance, motion, and time.

From a purely logical point of view, we must leave these concepts undefined. However, being human and full of hope that some degree of communication can result, we will proceed to describe some of these concepts intuitively. We do this because intuition is, perhaps, the greatest asset any scientist can have.

We think of a particle as a small spherical object, in three dimensions, or a circular object, in two dimensions, of uniform weight, or mass, whose center is called its centroid. A particle's motion is described completely by the motion of its center. All larger bodies are made up of particles. By a force, we think of a push or a pull. Motion will be thought of physiologically as follows. Let Δx and Δt be positive numbers. On an X-axis, let $X_k = k\Delta x$, $k = 0,1,2,\ldots,m$, while on a T-axis, let $t_j = j\Delta t$, $j = 0,1,2,\ldots,n$, as shown in Fig. 3.2. For illustrative purposes, assume that a particle P has center C at X_0 when $t = t_0$, at X_3 when $t = t_1$, at X_7 when $t = t_2$, at X_8 when $t = t_3$, and at X_6 when $t = t_4$. Then the motion of P from X_0 to X_6 is viewed merely as C's being at X_0, X_3, X_7, X_8, and X_6 at the respective times t_0, t_1, t_2, t_3, and t_4. Thus, P's motion is conceived of as a sequence of "stills", which is, of course, biologically acceptable and realized in motion pictures, where motion is observed from a finite sequence of stills projected with sufficient rapidity and transmitted as retinal images to the brain. The passage of time can be thought of in terms of the uniform motion of a time particle of mass Δt.

Theoretically, we are now ready to define velocity and acceleration. At each time t_k, let particle P, which is in motion along an X-axis, be positioned, or located, at $x(t_k) = x_k$. Note that x_k is the directed distance of P from the origin 0. Then P's velocity $v(t_k) = v_k$ at each t_k is defined as the average rate of change of its position with respect to time, as measured by the formula

$$\frac{v_{k+1}+v_k}{2} = \frac{x_{k+1}-x_k}{\Delta t}. \tag{3.5}$$

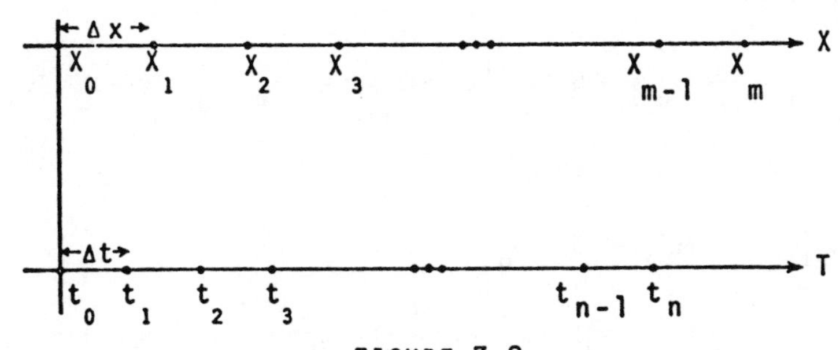

FIGURE 3.2

P's acceleration $a(t_k) = a_k$ at each t_k is defined as the average rate of change of its velocity with respect to time, as measured by the formula

$$a_k = \frac{v_{k+1}-v_k}{\Delta t}. \tag{3.6}$$

Note that formulas (3.5) and (3.6) were motivated by the discussion of gravity in Chapter II.

With regard to (3.5), observe that

$$\begin{aligned}
v_1 &= \frac{2}{\Delta t}[x_1 - x_0] - v_0 \\
v_2 &= \frac{2}{\Delta t}[x_2 - x_1] - v_1 = \frac{2}{\Delta t}[x_2 - 2x_1 + x_0] + v_0 \\
v_3 &= \frac{2}{\Delta t}[x_3 - x_2] - v_2 = \frac{2}{\Delta t}[x_3 - 2x_2 + 2x_1 - x_0] - v_0 \\
v_4 &= \frac{2}{\Delta t}[x_4 - x_3] - v_3 = \frac{2}{\Delta t}[x_4 - 2x_3 + 2x_2 - 2x_1 + x_0] + v_0, \\
&\vdots
\end{aligned}$$

which motivates the following theorem.

Theorem 3.2. Formula (3.5) implies that

$$v_1 = \frac{2}{\Delta t}[x_1 - x_0] - v_0$$

$$v_k = \frac{2}{\Delta t}[x_k + (-1)^k x_0 + 2\sum_{j=1}^{k-1}(-1)^j x_{k-j}] + (-1)^k v_0; \quad k \geq 2.$$

Proof. The proof of the theorem for $k = 1,2,3,4$ follows from the formulas which precede the statement of the theorem. Let us assume then that the theorem is true for $k = n$ and prove it is true for $k = n + 1$, that is, let us proceed by mathematial induction. We assume that

$$v_n = \frac{2}{\Delta t}[x_n + (-1)^n x_0 + 2\sum_{j=1}^{n-1}(-1)^j x_{n-j}] + (-1)^n v_0; \quad n \geq 2. \tag{3.7}$$

We wish to prove that

$$v_{n+1} = \frac{2}{\Delta t}[x_{n+1} + (-1)^{n+1} x_0 + 2\sum_{j=1}^{n}(-1)^j x_{n+1-j}] + (-1)^{n+1} v_0; \quad n \geq 2. \tag{3.8}$$

From (3.5), (3.7), and (3.8), then

$$\begin{aligned}
v_{n+1} &= \frac{2}{\Delta t}(x_{n+1}-x_n) - v_n \\
&= \frac{2}{\Delta t}(x_{n+1}-x_n) - \frac{2}{\Delta t}[x_n + (-1)^n x_0 + 2\sum_{j=1}^{n-1}(-1)^j x_{n-j}] - (-1)^n v_0
\end{aligned}$$

$$= \frac{2}{\Delta t}[x_{n+1} - (-1)^n x_0 - 2x_n - 2\sum_{j=1}^{n-1}(-1)^j x_{n-j}] - (-1)^n v_0$$

$$= \frac{2}{\Delta t}[x_{n+1} + (-1)^{n+1} x_0 + 2\sum_{j=1}^{n}(-1)^j x_{n+1-j}] + (-1)^{n+1} v_0,$$

d the theorem is proved.

an analogous fashion, study of $a_1, a_2, a_3, a_4, a_5, \ldots,$ motivates the following eorem.

eorem 3.3. If a_k is defined by (3.6), then

$$a_0 = \frac{2}{(\Delta t)^2}[x_1 - x_0 - v_0 \Delta t]$$

$$a_1 = \frac{2}{(\Delta t)^2}[x_2 - 3x_1 + 2x_0 + v_0 \Delta t]$$

$$a_{k-1} = \frac{2}{(\Delta t)^2}\left\{x_k - 3x_{k-1} + 2(-1)^k x_0 + 4\sum_{j=2}^{k-1}[(-1)^j x_{k-j}] + (-1)^k v_0 \Delta t\right\}, \quad k \geq 3.$$

oof. From (3.6) and Theorem 3.2,

$$a_0 = \frac{v_1 - v_0}{\Delta t} = \frac{1}{\Delta t}\left\{\frac{2}{\Delta t}(x_1 - x_0) - 2v_0\right\} = \frac{2}{(\Delta t)^2}[x_1 - x_0 - v_0 \Delta t]$$

$$a_1 = \frac{v_2 - v_1}{\Delta t} = \frac{1}{\Delta t}\left\{[\frac{2}{\Delta t}(x_2 - 2x_1 + x_0) + v_0] - [\frac{2}{\Delta t}(x_1 - x_0) - v_0]\right\}$$

$$= \frac{2}{(\Delta t)^2}[x_2 - 3x_1 + 2x_0 + v_0 \Delta t],$$

ile, for $k \geq 3$,

$$a_{k-1} = \frac{v_k - v_{k-1}}{\Delta t}$$

$$= \frac{1}{\Delta t}\left\{\frac{2}{\Delta t}[x_k + (-1)^k x_0 + 2\sum_{j=1}^{k-1}(-1)^j x_{k-j}] + (-1)^k v_0\right.$$

$$\left. - \frac{2}{\Delta t}[x_{k-1} + (-1)^{k-1} x_0 + 2\sum_{j=1}^{k-2}(-1)^j x_{k-j-1}] - (-1)^{k-1} v_0\right\}$$

$$= \frac{2}{(\Delta t)^2}\left\{x_k - 3x_{k-1} + 2(-1)^k x_0 + 4\sum_{j=2}^{k-1}[(-1)^j x_{k-j}] + (-1)^k v_0 \Delta t\right\},$$

d the theorem is proved.

will not have occasion to use Theorems 3.2 and 3.3 until Chapter XII. But, at at time, they will prove to be of exceptional value.

We turn finally to the question of which axioms, or assumptions, should be chosen to guide the development of our physical theory. Thus far, our experience with gravity suggests only that we should relate force and acceleration by a formula like (2.24). So, our first assumption is as follows.

Axiom 1. At each time t_k, the force $F_k = F(x_k,v_k,v_{k+1})$ acting in particle P and the acceleration a_k of P are related by the equation

$$ma_k = F(x_k,v_k,v_{k+1}), \tag{3.9}$$

where m is the mass of P.

Axiom 1 is only one of three basic "laws" which were assumed by Newton in his development of Mechanics. A second will be given in Section 3.4. The third will be given in Section 7.10, where its content will have special significance.

Note that if (3.9) is written in the equivalent form

$$\frac{v_{k+1}-v_k}{\Delta t} = \frac{F(x_k,v_k,v_{k+1})}{m}, \tag{3.10}$$

then we see that (3.9) is, in fact, a difference equation in v, from which arose our original interest in difference equations. Also, (3.9) is often called Newton's dynamical equation, for we shall show shortly how, when written in the equivalent form (3.10), this equation can be used to determine the motion of a particle from fixed initial conditions.

Note, also, that we are now in a position to prove a rather general and interesting theorem. This theorem is motivated by the observation that nowhere in the derivation of (2.28) was it ever necessary that F_k had to be the force of gravity.

Theorem 3.4. For $\Delta t > 0$, let $t_k = k\Delta t$. Let particle P, of mass m, be in motion along an X-axis. At time t_k, let P be located at $x_k = x(t_k)$, have velocity $v_k = v(t_k)$, and have acceleration $a_k = a(t_k)$. Let the kinetic energy K_k of P at t_k be defined by

$$K_k = \frac{1}{2}mv_k^2. \tag{3.11}$$

If

$$W_n = \sum_{k=0}^{n-1}(x_{k+1}-x_k)F_k, \quad n = 1,2,3,\ldots, \tag{3.12}$$

then

$$W_n = K_n - K_0. \tag{3.13}$$

Proof. Using (3.5), (3.6) and (3.9), the proof follows in the same way as (2.28) followed from (2.25).

We remark now that the product $(x_{k+1}-x_k)F_k$, that is, the product of the force a times the distance over which it acts, is taken as a measure of the "work" done F from time t_k to t_{k+1}. Thus, the total work W_n done by F from t_0 to is given by (2.25). The physical meaning of Theorem 3.4 is that the total work done by F from time t_0 to time t_n is always the difference of the kinetic ergies K_n and K_0.

3.4 REMARKS

rough the concept of acceleration, Axiom 1 relates a particle's motion to the rce acting on the particle. More naive questions about a particle's motion when force is acting on it were first answered by Galileo in his Principle of Inertia. is principle was, indeed, assumed to be so fundamental by Newton, that he took it another of his basic physical assumptions, and for completeness, is given now:

iom 2. If no force acts on a particle and the particle is at rest, then it will main at rest, while if it is in motion, then the motion will continue along a raight line and at a constant velocity.

r future purposes, note also that it will always be necessary to set up force rmulas which are consistent with (3.9), that is, the force must be positive when e acceleration is positive and the force must be negative when the acceleration negative. Remembering this rule will enable one to understand quite often why rtain formulas are preceded by negative signs.

3.5 EXERCISES - CHAPTER 3

.1 List all the points of the R_{n+1} set in which :

(a) $a = 0, \quad \Delta x = 0.25, \quad n = 5$

(b) $a = 1, \quad \Delta x = 0.25, \quad n = 6$

(c) $a = -1, \quad \Delta x = 0.25, \quad n = 7$

(d) $a = -1, \quad \Delta x = 0.3, \quad n = 7$

(e) $a = 0.7, \quad \Delta x = 0.2, \quad n = 13$

(f) $a = 0, \quad \Delta x = 0.1, \quad n = 100$

.2 Describe the R_{n+1} set which divides each of the following intervals $a \leq x \leq b$ into the given number of n equal parts.

(a) $a = 0, \quad b = 1, \quad n = 10$

(b) $a = 0, \quad b = 2, \quad n = 10$

(c) $a = 0, \quad b = 7, \quad n = 10$

(d) $a = -1, \quad b = 3, \quad n = 10$

(e) $a = 0, \quad b = 2, \quad n = 100$

(f) $a = 0$, $b = 3$, $n = 37$

(g) $a = -3$, $b = 5$, $n = 43$

(h) $a = 0.2$, $b = 1.7$, $n = 100$

(i) $a = 0.2$, $b = 1.7$, $n = 62$

(j) $a = 0$, $b = 3$, $n = 1000$

(k) $a = 0$, $b = 3.3$, $n = 10000$.

3.3 On the R_5 set $x_0 = 0$, $x_1 = 0.25$, $x_2 = 0.50$, $x_3 = 0.75$, $x_4 = 1.00$, graph each of the following discrete functions.

(a) $y_k = x_k^2$, $k = 0,1,2,3,4$

(b) $y_k = \sin \frac{\pi x_k}{4}$, $k = 0,1,2,3,4$

(c) $y_i = 1 - 2x_i + x_i^3$, $i = 0,1,2,3,4$

(d) $y_j = e^{-x_j}$, $j = 0,1,2,3,4$.

3.4 On the R_{21} set $x_k = -1 + \frac{k}{10}$, $k = 0,1,2,\ldots,20$, graph each of the following discrete functions.

(a) $y_k = x_k^2$

(b) $y_k = \sin \frac{\pi x_k}{4}$

(c) $y_k = 1 - 2x_k + x_k^3$.

3.5 On the R_{101} set $x_k = \frac{k}{100}$, $k = 0,1,2,\ldots,100$, graph each of the following discrete functions.

(a) $y_k = 2x_k$

(b) $y_k = -x_k^2$

(c) $y_k = \frac{1}{1+x_k}$.

3.6 On the R_{1001} set $x_k = -1 + \frac{k}{500}$, $k = 0,1,2,\ldots,1000$, graph each of the following discrete functions.

(a) $y_k = -2x_k + 1$

(b) $y_k = \sin \pi x_k$

(c) $y_k = -3 \cos \pi x_k$

(d) $y_k = 1 - x_k^3$.

.7 On R_{n+1} set $x_0, x_1, \ldots, x_n$, which of the following are difference equations? Of those which are difference equations, which are explicit and which are implicit?

(a) $y_{k+1} = y_k^2 - 3x_k$, $k = 0,1,2,\ldots,n-1$

(b) $y_{k+1} = y_k^2 - 3x_k + 2$, $k = 0,1,2,\ldots,n-1$

(c) $y_{k+1} = -3x_k + 2$, $k = 0,1,2,\ldots,n-1$

(d) $y_{k+1} = 2$, $k = 0,1,2,\ldots,n-1$

(e) $y_{k+1} = 2 \sin x_k$, $k = 0,1,2,\ldots,n-1$

(f) $y_{k+1} = 2 \sin y_k$, $k = 0,1,2,\ldots,n-1$

(g) $y_{k+1} = 2 \sin y_{k+1}$, $k = 0,1,2,\ldots,n-1$

(h) $y_{k+1} = x_k + 2 \sin y_{k+1}$, $k = 0,1,2,\ldots,n-1$

(i) $y_{k+1} = x_k + y_k + 2 \sin y_{k+1}$, $k = 0,1,2,\ldots,n-1$

(j) $y_{k+1} = x_k + 2 \sin y_k - 2 \sin y_{k+1}$, $k = 0,1,2,\ldots,n-1$

(k) $y_{k+1} = x_k + 2 \sin(y_k - y_{k+1})$, $k = 0,1,2,\ldots,n-1$.

8 On the R_5 set $x_0 = 0$, $x_1 = 0.25$, $x_2 = 0.50$, $x_3 = 0.75$, $x_4 = 1.00$, find three solutions of each of the following explicit difference equations. Check your answers.

(a) $y_{k+1} = -2y_k$, $k = 0,1,2,3$

(b) $y_{k+1} = y_k + \frac{x_k}{10}$, $k = 0,1,2,3$

(c) $y_{k+1} = y_k + 3 - x_k^2$, $k = 0,1,2,3$

(d) $y_{k+1} = y_k/(1+x_k^2)$, $k = 0,1,2,3$

(e) $y_{k+1} = y_k^2/(1+x_k^2)$, $k = 0,1,2,3$.

.9 On the R_5 set given in Exercise 3.8, find the unique solution of each difference equation if $y_0 = 1$. Check your answer.

.10 On the R_{11} set $x_k = \frac{k}{10}$, $k = 0,1,2,\ldots,10$, find and graph the solution of the initial value problem defined by

$$y_{k+1} = -2y_k, \quad y_0 = 1.$$

Check your answer.

3.11 On the R_{21} set $x_k = 1 + \frac{k}{10}$, $k = 0,1,2,\ldots,20$, find and graph the solution of the initial value problem

$$y_{k+1} = y_k - \frac{x_k^3}{100}, \quad y_0 = 0.$$

Check your answer.

3.12 On the R_{101} set $x_k = \frac{k}{100}$, $k = 0,1,2,\ldots,100$, find and graph the solution of the initial value problem

$$y_{k+1} = \frac{1-x_k}{1+y_k^2}, \quad y_0 = -1.$$

Check your answer.

3.13 On the R_{2001} set $x_k = -1 + \frac{k}{1000}$, $k = 0,1,2,\ldots,2000$, find and graph the solution of the initial value problem

$$y_{k+1} = y_k - \frac{x_k^2}{1000}, \quad y_0 = 1.$$

Check your answer.

3.14 On the R_{1001} set $x_k = \frac{k}{1000}$, $k = 0,1,2,3,\ldots,1000$, find and graph the solution of the initial value problem

$$y_{k+1} = y_k + \frac{1}{(1+x_k)^5}.$$

3.15 Show, in detail, with regard to the proof of Theorem 3.2, that

$$-2x_n - 2\sum_{j=1}^{n-1} (-1)^j x_{n-j} = 2\sum_{j=1}^{n} (-1)^j x_{n+1-j}.$$

3.16 Prove that for any integer m:

$$\sum_{j=0}^{n} f(j) = \sum_{j=m}^{n+m} f(j-m).$$

3.17 Show that the force of gravity acting on a particle P of mass m is given by $F = -32m$, where, if m is measured in pounds, the units of F are ft.lbs/sec^2.

3.18 Discuss the following free translation of Zeno's "Achilles and the Tortoise" paradox.

A fast runner and a slow tortoise are to have a race. Because of the runner's superior speed, the tortoise is allowed to begin the race at a positive distance d ahead of the runner (see the figure). Let the runner's initial

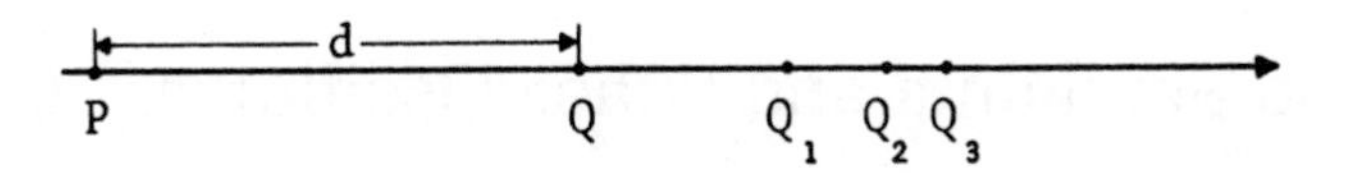

point be P and that of the tortoise be Q. After the race has begun, the runner must reach the point Q, which takes time, during which the tortoise moves ahead to a new point Q_1. The runner must then reach the point Q_1, which takes time, during which the tortoise moves ahead to a new point Q_2. The runner must then reach the point Q_2, which takes time, during which the tortoise moves ahead to a new point Q_3, and so forth. Thus the runner must always reach a point where the tortoise has already been, from which it follows that the runner, no matter what his speed, can never overtake the tortoise.

CHAPTER 4

The Pendulum and Other Oscillators

4.1 INTRODUCTION

With a little thought, one realizes that gravity is amenable to various interesting experiments which can be performed rather easily. For example, one could try to constrain the motion of a falling particle by, for example, tying one end of a string to the particle and the other end to a fixed hook. The resulting motion, though interesting to observe, may be very complex. Further interest can be added by using an elastic string in the above experiment, but the resulting motion still may be exceptionally complex. However, if in place of the string one were to use a light bar, which is attached to the particle and is hinged on the ceiling, the resulting configuration is called a pendulum and the motion simplifies considerably. The two gross effects which are immediately observable are that the particle's motion appears to be on the arc of a circle, and that the motion decays and finally ceases, or, more technically, damps out. One could oil the hinge or introduce a bearing system to produce more motion, but in all cases friction cannot be eliminated entirely, and damping results. This is true, incidentally, even if the pendulum is placed in a vacuum jar.

There is a third effect in pendulum motion which also is easily observable, but which requires the use of a clock to make various time measurements. This effect is that the time between successive swings of a pendulum decreases. More precisely, if τ_k is the time of any pendulum swing in, say, the clockwise direction, while τ_{k+1} is the time of the very next pendulum swing, which must be in the counterclockwise direction, then $\tau_{k+1} < \tau_k$ for any k.

With the above experimental results as a guide, let us try now to formulate and study a theoretical model of pendulum motion.

4.2 A THEORETICAL PENDULUM

Consider mutually perpendicular coordinate axes V and H, as shown in Fig. 4.1, where V is the vertical axis and H the horizontal. Let $AO'B$ be a semicircle, of radius ℓ, in the third and fourth quadrants and let P be a particle of mass m whose center is constrained to move on the semicircle. If P is released from a position of rest and is acted upon by the force of gravity, we say that the motion of P simulates the motion of a pendulum.

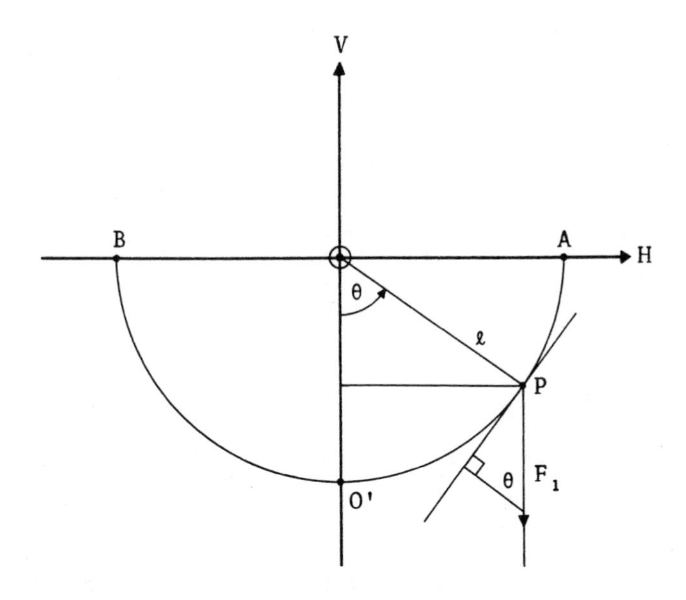

FIGURE 4.1

In the case of a real pendulum, the radial arm OP could have a significant weight. In our model, we are assuming that, for all practical purposes, this weight is negligible when compared to the weight of P. Also, our assumption that P's motion is on a circular arc implies that any "bend" which may develop during the motion of the arm OP is also negligible when compared with other effects. We hope, as we make these assumptions, that we have not thrown away something of importance, and if our theoretical results do not agree with experimental results, we will have to reconsider the formulation.

Now, of the three experimental results described in Section 4.1, the first is already satisfied by our assumption that motion is on a circular arc. Let us see then how to continue in a manner consistent with the other two.

We want next to study, say, the damping effect, but this presents a minor problem. The amount of friction generated will depend on the <u>velocity</u> of P. From Fig. 4.1, it is apparent that P's motion is two dimensional. Our problem is that we developed the concept of velocity in only one dimension. Nevertheless, the assumption that P's motion is on a circular arc enables us to proceed with the analysis in one dimension as follows. Consider a new coordinate system X, which is one dimensional, and which results from the bending of circular arc BO'A in Fig. 4.1 into the straight line segment BO'A shown in Fig. 4.2. We take O' as the origin of

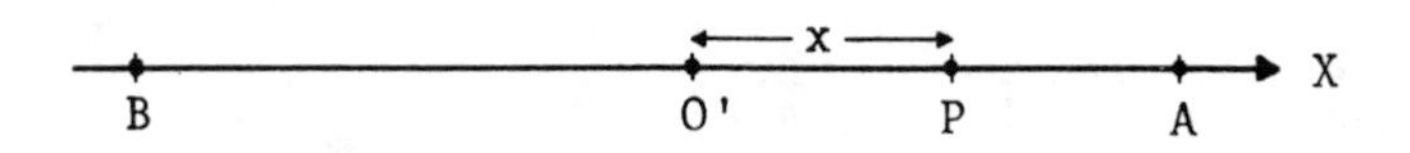

FIGURE 4.2

this axis. If the directed length of the circular arc $O'P$ in Fig. 4.1 is x, then we take this to be the X-coordinate of P on the X-axis of Fig. 4.2. To determine the motion of P on this axis, we must determine the actual forces in action, and for this purpose we utilize the arrangement of Fig. 4.1 first.

The actual force acting on P is assumed to consist of two parts, F_1 due to gravity and F_2 due to friction. The gravitational force acting on P is $-32m$ in ft-second units, and is shown to be acting vertically in Fig. 4.1. But, since P is constrained to move on a circular arc, it follows from the similar triangle of Fig. 4.1 that only $-32m \sin \theta$ of this force actually acts on P in the direction of interest. Moreover, assuming that θ is measured in radians, so that $x = \ell\theta$, it follows that

$$F_1 = -32m \sin \frac{x}{\ell}.$$

The question of how to define F_2, the force on P due to friction, is very difficult to answer. It is reasonable, from a very crude point of view, simply to assume that F_2 varies directly with the velocity of P, that is, the faster P moves, the more frictional force is generated, the magnitude of which is linear with respect to the speed. So, for simplicity, we will assume that

$$F_2 = -\alpha v, \quad \alpha > 0$$

where α is a positive "friction" constant and the sign of F_2 is chosen to be negative to make it work <u>against</u> the motion. We assume, finally, then, that in P's motion on the circular arc of Fig. 4.1, the force F which contributes to this motion is the sum of the gravitational and frictional components, so that

$$F = F_1 + F_2,$$

or, more precisely,

$$F = -32m \sin \frac{x}{\ell} - \alpha v. \tag{4.1}$$

It is this force that we consider to be acting on P when we analyze its motion on the X-axis of Fig. 4.2.

Consider then the motion of P on the X-axis of Fig. 4.2. For $\Delta t > 0$, set $t_k = k\Delta t$, $k = 0,1,2,\ldots$. From (4.1), assume that at each t_k, the force acting on P is

$$F_k = -32m \sin \frac{x_k}{\ell} - \alpha v_k. \tag{4.2}$$

Then, (4.2) together with (2.5), (2.6) and (2.9) imply

$$\frac{v_{k+1}-v_k}{\Delta t} = -32 \sin \frac{x_k}{\ell} - \frac{\alpha}{m} v_k \tag{4.3}$$

$$\frac{x_{k+1}-x_k}{\Delta t} = \frac{v_{k+1}+v_k}{2} \tag{4.4}$$

Now, interestingly enough, (4.3) and (4.4) can be rewritten as

$$v_{k+1} = v_k + \Delta t(-32 \sin \frac{x_k}{\ell} - \frac{\alpha}{m} v_k) \tag{4.3'}$$

$$x_{k+1} = x_k + \frac{\Delta t}{2}(v_{k+1}+v_k) \tag{4.4'}$$

and these are very useful, explicit recursion formulas. Indeed, if one knows the initial data x_0 and v_0, then (4.3') and (4.4') yield, recursively, $v_1, x_1, v_2, x_2, v_3, x_3, \ldots$, and so forth. For illustrative purposes, then, let us consider next an actual computer example in which the iteration is performed and the resulting pendulum motion is analyzed.

Consider a pendulum for which $\ell = 1$, $m = 1$, and $\alpha = 1.6$. Drop the pendulum from a position of rest from a 45° position, so that $x_0 = \frac{\pi}{4}$, $v_0 = 0$. Finally, for the actual computations, let $\Delta t = 0.001$. Then (4.3') and (4.4') reduce to

$$v_{k+1} = v_k + (0.001)(-32 \sin x_k - (1.6)v_k) \tag{4.3''}$$

$$x_{k+1} = x_k + (0.0005)(v_{k+1}+v_k). \tag{4.4''}$$

Beginning with $x_0 = \frac{\pi}{4}$, $v_0 = 0$, (4.3") and (4.4") were iterated up to $k = 15000$ on the UNIVAC 1108 in under 30 seconds. A plot of each position x_k at time t_k up to t_{3391} is shown in Fig. 4.3. The graph shows strong damping and peak values of 0.78540, -0.49247, 0.31305, -0.20003, 0.12808, -0.08208 and 0.05262 at the respective times 0, 0.577, 1.144, 1.708, 2.269, 2.830 and 3.391. Simple calculations then reveal that the time required for the particle to travel from one peak successively to the next decreased monotonically, which is consistent with the third observation made from physical experimentation, that is, the time between successive swings is decreasing. Thus, this particular computer example of our model agrees in every respect with the physical observations made in Section 4.1.

One can now vary the parameters in (4.3') and (4.4') to generate qualitatively different pendulum motions. Thus, for example, decreasing α yields less damping, and so forth. However, such computer experimentation reveals, quickly enough, that for certain parameter choices of α and Δt, one has computational instability. We want to avoid these as being nonrepresentative of physical pendula, whose motions are known to be exceptionally stable. So, to discover which parameter combinations to avoid, we proceed as follows. Let us fix $\ell = 1$, $m = 1$, $x_0 = \frac{\pi}{4}$, $v_0 = 0$. Consider then $\alpha = 0.8$ and iterate (4.3') and (4.4') with each of $\Delta t = 0.025$, 0.050, 0.075, 0.100,...,0.500. The resulting computations are stable for $\Delta t = 0.025$, and are unstable for $\Delta t = 0.050$, 0.075, 0.010,...,0.500. Now repeat the above calculations but with $\alpha = 1.6$. Then the resulting computations are stable for $\Delta t = 0.025$, 0.050, 0.075, and are unstable for $\Delta t = 0.100$, 0.125, 0.150, 0.175,...,0.500. Continue the computations with $\alpha = 2.4$, then $\alpha = 3.2$, and $\alpha = 4.0$. For each of the values of α considered, determine the first value of the Δt sequence for which instability occurs. For $\alpha = 0.8$, this would be $\Delta t = 0.050$, for $\alpha = 1.6$, this would be $\Delta t = 0.100$, and so forth. Plot these values of Δt against α, as shown in Fig. 4.4. Remarkably enough, the plotted points all lie on the straight line

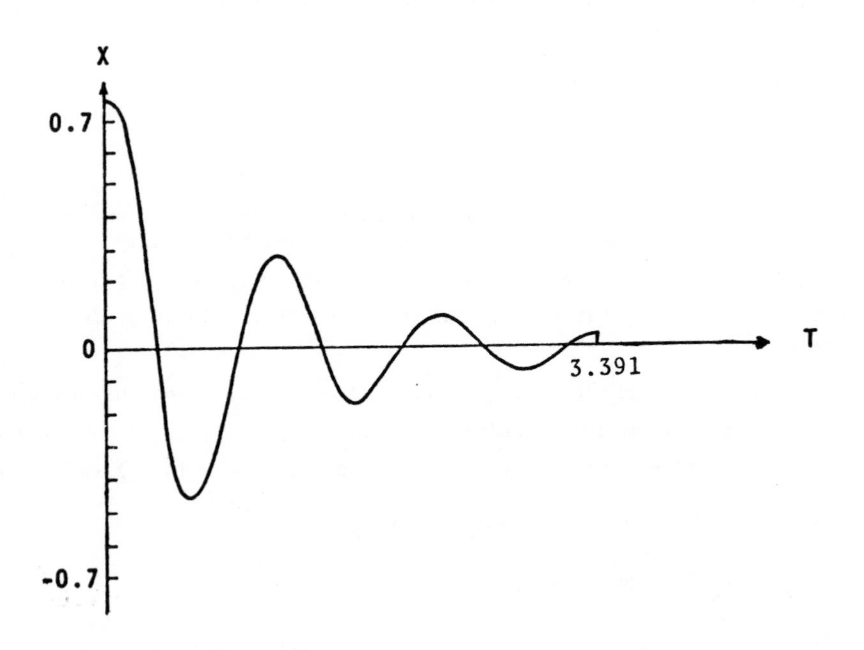

FIGURE 4.3

$$\Delta t = \frac{\alpha}{16}.$$

One can now repeat the calculation with new choices for x_0 and v_0, but still the same relationship follows. Thus, the computations suggest that for all α, a necessary condition for stability is that

$$\Delta t < \frac{\alpha}{16}, \tag{4.5}$$

which is, indeed, correct (see Greenspan (1973(a)).

Note that, throughout our discussion, we assumed that damping was an unavoidable facet of the actual motion of a pendulum. The above stability condition reflects this assumption in that, if one were to violate it by setting $\alpha = 0$, then, indeed no positive Δt would exist to yield stable computations.

4.3 THE HARMONIC OSCILLATOR

A pendulum is a special example of a large number of other physical systems called oscillators. More precisely, oscillation is motion back and forth over all, or part, of a straight or a curved path. Vibrating springs and even electrons are often studied from the oscillator point of view. But, in the case of an electron, for example, the damping effect seems to be so small that usually it is neglected. Such an oscillator is called a harmonic oscillator and we will study it next. It

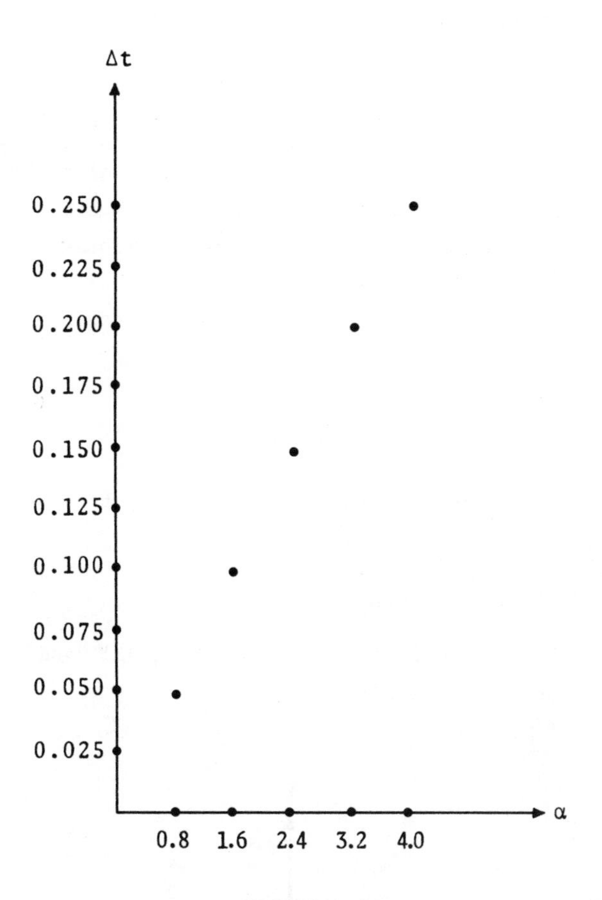

FIGURE 4.4

needs special attention because, as indicated at the end of Section 4.3, the stability condition derived for damped oscillators does not allow one simply to set $\alpha = 0$. We will concentrate then on how to formulate the equations of motion of such an undamped, or, energy conserving, oscillator. The analysis will be given only for motion along an X-axis, since, as shown in Section 4.2, motion on a circular path, for example, also can be analyzed in terms of motion along an appropriate X-axis.

For simplicity, let us examine (4.1) again. To simulate the motion of a harmonic oscillator, we first neglect the damping term $-\alpha v$. Next, since we do not wish to consider only gravity, we replace the coefficient $-32m$ by a parameter $-w$, where $w > 0$. Finally, if one were to think of the motion of an electron in an atom of a solid, this motion would be very small. For this reason, we merely set $\ell = 1$ and consider $|x|$ to be much smaller than unity, which is denoted by the double use of the "less than" sign $<$ as follows:

$$|x| << 1. \qquad (4.6)$$

Thus, (4.1) takes the form

$$F = -w \sin x, \quad w > 0, \tag{4.7}$$

where x is to be in the range defined by (4.6).

Now if x is to be restricted by (4.6), then one can make a highly accurate approximation for $\sin x$ in (4.7) as follows. Let C be a circle of unit radius and center O, as shown in Fig. 4.5. Construct a central angle ϕ which is in the range $0 < \phi < \frac{\pi}{2}$ and which subtends an arc PN. Let the perpendicular from P to ON meet ON in the point M. Let the perpendicular at N to ON meet OP (extended) in the point R. Then

$$0 < \text{area } \Delta OPM < \text{area section } OPN < \text{area } \Delta ORN. \tag{4.8}$$

or,

$$0 < \frac{1}{2}\sin\phi\cos\phi < \frac{1}{2}\phi < \frac{1}{2}\tan\phi \tag{4.9}$$

where, of course, ϕ is measured in radians. Hence, since $0 < \phi < \frac{\pi}{2}$, (4.9) implies

$$0 < \cos\phi < \frac{\phi}{\sin\phi} < \frac{1}{\cos\phi}. \tag{4.10}$$

But, when ϕ is small, that is, $\phi \sim 0$, then $\cos\phi \sim 1$ and $\frac{1}{\cos\phi} \sim 1$, so that (4.10) yields for ϕ small that

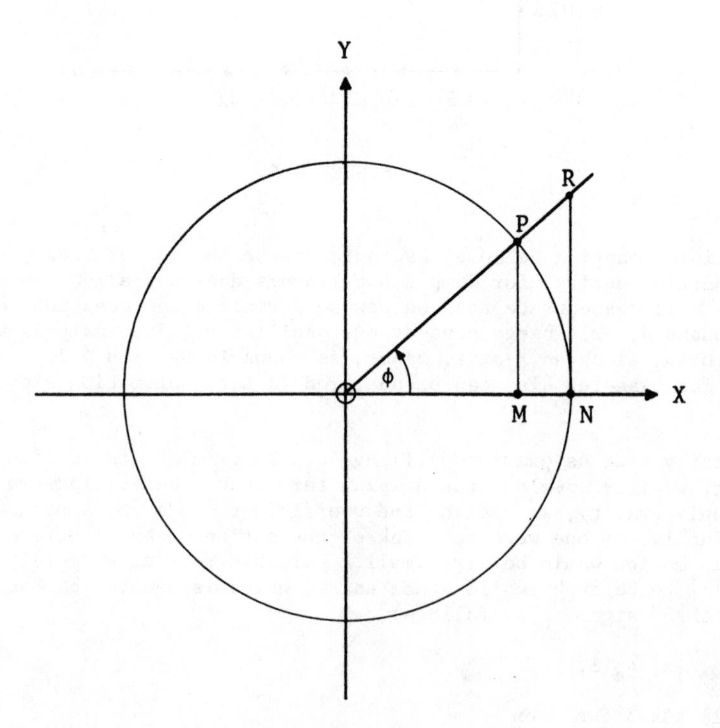

FIGURE 4.5

$$\phi \sim \sin \phi \tag{4.11}$$

A similar argument for $\phi < 0$ also implies (4.11).

To return now to (4.7), the result (4.11) allows us to use the approximation

$$F = -wx, \tag{4.12}$$

or, more particularly,

$$F_k = -wx_k. \tag{4.13}$$

Substitution of (4.13) into (3.12) yields then

$$W_n = -\sum_{k=0}^{n-1}(x_{k+1}-x_k)wx_k = -w\sum_{k=0}^{n-1}(x_{k+1}x_k-x_k^2). \tag{4.14}$$

But if we wish, as in the case of gravity, to develop a formula for potential energy V_k such that

$$W_n = V_n - V_0, \tag{4.15}$$

then clearly (4.13) is not it, because, to get (4.15), one would require that (4.14) be telescopic, so as to yield the difference on the right-hand side of (4.15). So, we must reexamine (4.13) and ask how to reformulate it such that W_n is telescopic. The answer, with a little thought, is quite easy. Simply choose

$$F_k = -w\,\frac{x_{k+1}+x_k}{2}, \tag{4.16}$$

for then

$$W_n = \sum_{k=0}^{n-1}(x_{k+1}-x_k)F_k = -\frac{w}{2}\sum_{k=0}^{n-1}(x_{k+1}-x_k)(x_{k+1}+x_k)$$

$$= -\frac{w}{2}\sum_{k=0}^{n-1}(x_{k+1}^2-x_k^2) = -\frac{w}{2}x_n^2 + \frac{w}{2}x_0^2.$$

Defining the potential energy of P at t_k by $V_k = \frac{w}{2}x_k^2$ yields

$$W_n = -V_n + V_0, \quad n = 1,2,3,4,\ldots\ . \tag{4.17}$$

Finally, elimination of W_n between (3.13) and (4.17) yields

$$K_n + V_n = K_0 + V_0, \quad n = 1,2,3,4,5,\ldots \tag{4.18}$$

which is the law of conservation of energy.

Thus, all is well if the harmonic oscillator is defined by

$$ma_k = F_k, \tag{4.19}$$

where

$$F_k = -w\frac{x_{k+1}+x_k}{2}, \quad w > 0 \tag{4.20}$$

4.4 HARMONIC MOTION

The motion of a harmonic oscillator is called harmonic motion. Such motion can be studied on a computer in the following way. From (3.6), (4.19) and (4.20), one has

$$\frac{v_{k+1}-v_k}{\Delta t} = -\frac{w}{m}\frac{x_{k+1}+x_k}{2}, \tag{4.21}$$

which, coupled with (3.5):

$$\frac{v_{k+1}+v_k}{2} = \frac{x_{k+1}-x_k}{\Delta t}, \tag{4.22}$$

form a system of two linear algebraic equations for x_{k+1} and v_{k+1} in terms of x_k and v_k. Rewriting this system in the form

$$\frac{w\Delta t}{2m}x_{k+1} + v_{k+1} = -\frac{w\Delta t}{2m}x_k + v_k \tag{4.23}$$

$$\frac{2}{\Delta t}x_{k+1} - v_{k+1} = \frac{2}{\Delta t}x_k + v_k, \tag{4.24}$$

one finds that the unique solution is

$$x_{k+1} = \frac{4m-w\Delta t^2}{4m+w\Delta t^2}x_k + \frac{4m\Delta t}{4m+w\Delta t^2}v_k \tag{4.25}$$

$$v_{k+1} = -\frac{4w\Delta t}{4m+w\Delta t^2}x_k + \frac{4m-w\Delta t^2}{4m+w\Delta t^2}v_k. \tag{4.26}$$

Thus, (4.25) and (4.26) are the recursion formulas from which harmonic motion can be analyzed once x_0, v_0, m, w, and Δt are given.

As a particular example, let us take $x_0 = \frac{\pi}{4}$, $v_0 = 0$, $m = w = 1$ and $\Delta t = 10^{-3}$. The iteration was executed up to $k = 150{,}000$ and the output is shown, typically, up to $k = 12566$ in Fig. 4.6. Observe that the graph appears to be a cosine curve whose amplitude is 0.78540 and whose period, rounded to two decimal places, is 6.28. Thus, the motion is periodic and, as was to be expected, undamped, or, conservative. These characteristics are, in fact, common to all harmonic oscillators.

As a computational check on the conservative nature of (4.25) and (4.26), Δt, in the above example, was increased to 0.1. The numerical results were essentially the same as those described above. The FORTRAN program for this particular run is given in Appendix A.

Finally, let us show that the simplistic nature of any harmonic oscillator allows us, with some mathematical trickery, to analyze its motion completely _without_ the

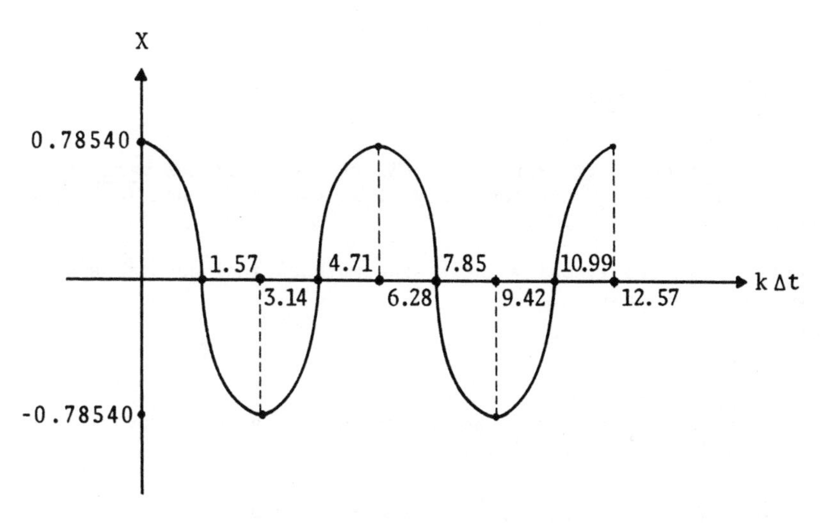

FIGURE 4.6

use of a computer. We will do this for the illustrative example given above, but the discussion extends readily to any other harmonic oscillator.

For $m = w = 1$, and for <u>any</u> Δt, (4.21) and (4.22) can be rewritten as

$$-v_{k+1} + v_k = \frac{\Delta t}{2}(x_{k+1}+x_k) \tag{4.27}$$

$$v_{k+1} + v_k = \frac{2}{\Delta t}(x_{k+1}-x_k) \tag{4.28}$$

which, by addition and simplification, imply

$$v_k = \left(\frac{\Delta t^2+4}{4\Delta t}\right)x_{k+1} + \left(\frac{\Delta t^2-4}{4\Delta t}\right)x_k. \tag{4.29}$$

But (4.29) must be valid for all k, so that it implies

$$v_{k+1} = \left(\frac{\Delta t^2+4}{4\Delta t}\right)x_{k+2} + \left(\frac{\Delta t^2-4}{4\Delta t}\right)x_{k+1}. \tag{4.30}$$

Addition of (4.29) and (4.30) yields $v_{k+1} + v_k$ in terms of x_{k+2}, x_{k+1}, and x_k, which, when substituted into the left-hand side of (4.28), implies

$$(\Delta t^2+4)x_{k+2} + 2(\Delta t^2-4)x_{k+1} + (\Delta t^2+4)x_k = 0. \tag{4.31}$$

Formula (4.31) is another recursion formula for x, but to use it we would have to know x_0 and x_1. From the initial data, however,

$$x_0 = \frac{\pi}{4}, \tag{4.32}$$

and, since $v_0 = 0$, (4.25) implies

$$x_1 = \frac{\pi}{4}\left(\frac{4-\Delta t^2}{4+\Delta t^2}\right). \tag{4.33}$$

Note that if we wanted to generate the oscillator's positions without also generating its velocities, which are necessary if we use (4.25) and (4.26), we could do this by means of (4.31) - (4.33).

Let us now see if we can actually guess some mathematical solutions of (4.31). For example, let us guess that

$$x_k = \lambda^k, \tag{4.34}$$

where λ is a nonzero constant to be determined, is a solution. Then direct substitution of (4.34) into (4.31) implies

$$(\Delta t^2+4)\lambda^{k+2} + 2(\Delta t^2-4)\lambda^{k+1} + (\Delta t^2+4)\lambda^k = 0. \tag{4.35}$$

Since λ has been assumed to be a nonzero constant, (4.35) implies

$$(\Delta t^2+4)\lambda^2 + 2(\Delta t^2-4)\lambda + (\Delta t^2+4) = 0$$

from which we find the two complex conjugate roots

$$\lambda_1 = \frac{4-\Delta t^2}{4+\Delta t^2} + i\frac{4\Delta t}{4+\Delta t^2} \tag{4.36}$$

$$\lambda_2 = \frac{4-\Delta t^2}{4+\Delta t^2} - i\frac{4\Delta t}{4+\Delta t^2}. \tag{4.37}$$

Now, by (4.34), we are going to be interested in λ_1^k and λ_2^k. These quantities would be much easier to handle if λ_1 and λ_2 were in polar form, since

$$[r(\cos\theta + i\sin\theta)]^n \equiv r^n(\cos n\theta + i\sin n\theta).$$

To put λ_1, then, in polar form, note that

$$|\lambda_1| = \left[\left(\frac{4-\Delta t^2}{4+\Delta t^2}\right)^2 + \left(\frac{4\Delta t}{4+\Delta t^2}\right)^2\right]^{1/2} = 1.$$

Since λ_2 is the complex conjugate of λ_1, then

$$|\lambda_1| = |\lambda_2| = 1 \tag{4.38}$$

and

$$\lambda_1 = \cos\theta + i\sin\theta, \quad \lambda_2 = \cos\theta - i\sin\theta,$$

where

$$\theta = \cos^{-1} \frac{4-\Delta t^2}{4+\Delta t^2}. \tag{4.39}$$

We have now two possible choices for x_k, namely, by (4.34),

$$\begin{cases} x_k = (\cos\theta + i \sin\theta)^k = \cos k\theta + i \sin k\theta \\ x_k = (\cos\theta - i \sin\theta)^k = \cos k\theta - i \sin k\theta, \end{cases} \tag{4.40}$$

neither of which is very appealing because each is a complex number. Since all the mathematics of the formulation was real, it is desirable to have only real solutions. The question which naturally comes to mind is whether our real solutions are, somehow, submerged in these complex solutions. And if this is the case, how can we extract them from (4.40)?

Let us proceed on the assumption that the real solutions are, indeed, hidden in (4.40) and try to extract them. Intuitively, since (4.40) are complex conjugates, we might try multiplication in some fashion that will yield only real numbers. For example, consider

$$x_k = (a+bi)(\cos k\theta + i \sin k\theta) + (a-bi)(\cos k\theta - i \sin k\theta), \tag{4.41}$$

where a and b are real parameters. Then, expansion of (4.41) yields, remarkably enough,

$$x_k = 2a \cos k\theta - 2b \sin k\theta. \tag{4.42}$$

Thus, (4.42) presents itself as an extracted real solution, and it has two real parameters a and b. But now these can be determined from the known values x_0 and x_1 of (4.32) and (4.33), since, setting $k = 0$ and $k = 1$ in (4.42) then implies

$$\frac{\pi}{4} = 2a \cos 0 - 2b \sin 0 \tag{4.43}$$

$$\frac{\pi}{4}\left(\frac{4-\Delta t^2}{4+\Delta t^2}\right) = 2a \cos\theta - 2b \sin\theta. \tag{4.44}$$

From (4.43), one has $a = \frac{\pi}{8}$. From (4.39) and (4.44), one has then

$$\frac{\pi}{4}\left(\frac{4-\Delta t^2}{4+\Delta t^2}\right) = \frac{\pi}{4}\left(\frac{4-\Delta t^2}{4+\Delta t^2}\right) - 2b\left(\frac{4\Delta t}{4+\Delta t^2}\right),$$

which implies $b = 0$. Thus, finally, (4.42) yields

$$x_k = \frac{\pi}{4} \cos k\theta \tag{4.45}$$

which is periodic and has amplitude $\frac{\pi}{4}$, in agreement with the rounded computational results shown graphically in Fig. 4.6.

4.5 REMARKS

There are important oscillation problems which are very different from those studied in the previous sections. Usually, the equations for these problems are much more complicated than those of pendula and harmonic oscillators. Typical of these equations is

$$a_k = \lambda(1-x_k^2)v_k - x_k, \quad k = 0,1,..., \tag{4.46}$$

where λ is a nonnegative parameter. The motion associated with (4.46) is studie in advanced electrical circuitry. The associated oscillator is called the van der Pol oscillator (see, e.g., F. L. H. M. Stumpers). Interestingly enough, we are able to solve problems associated with (4.46) by the methods already developed (see Greenspan (1972)) and some of the simpler of these are given in the Exercises

4.6 EXERCISES - CHAPTER 4

4.1 Compare the motions of pendula which differ only in the lengths of their radial arms.

4.2 Graph and describe the pendulum motion generated by (4.3') and (4.4') for each of the following sets of parameters:

(a) $\ell = 1$, $m = 3$, $\alpha = 0.3$, $x_0 = \frac{\pi}{4}$, $v_0 = 0$, $\Delta t = 0.01$

(b) $\ell = 5$, $m = 3$, $\alpha = 0.3$, $x_0 = \frac{\pi}{4}$, $v_0 = 0$, $\Delta t = 0.01$

(c) $\ell = 5$, $m = 1$, $\alpha = 0.3$, $x_0 = \frac{\pi}{4}$, $v_0 = 0$, $\Delta t = 0.01$

(d) $\ell = 5$, $m = 1$, $\alpha = 0.3$, $x_0 = \pi$, $v_0 = 0$, $\Delta t = 0.01$

(e) $\ell = 5$, $m = 1$, $\alpha = 0.3$, $x_0 = \frac{\pi}{2}$, $v_0 = 0$, $\Delta t = 0.01$

(f) $\ell = 5$, $m = 1$, $\alpha = 0.3$, $x_0 = \frac{\pi}{2}$, $v_0 = -3$, $\Delta t = 0.01$

(g) $\ell = 5$, $m = 1$, $\alpha = 0.3$, $x_0 = \frac{\pi}{2}$, $v_0 = 3$, $\Delta t = 0.01$

(h) $\ell = 5$, $m = 1$, $\alpha = 0.1$, $x_0 = \frac{\pi}{2}$, $v_0 = 0$, $\Delta t = 0.01$

(i) $\ell = 5$, $m = 1$, $\alpha = 0.02$, $x_0 = \frac{\pi}{2}$, $v_0 = 0$, $\Delta t = 0.01$

(j) $\ell = 5$, $m = 1$, $\alpha = 0.005$, $x_0 = \frac{\pi}{2}$, $v_0 = 0$, $\Delta t = 0.01$.

4.3 Prove the identity

$$(\cos\theta + i\sin\theta)^k = \cos k\theta + i\sin k\theta, \quad k = 1,2,3,\ldots .$$

4.4 Describe the harmonic motion which results from each of the following sets o parameters:

(a) $w = 1$, $m = 1$, $x_0 = 0$, $v_0 = 0$, $\Delta t = 0.01$

(b) $w = 1, \quad m = 1, \quad x_0 = \pi, \quad v_0 = 0, \quad \Delta t = 0.01$

(c) $w = 1, \quad m = 1, \quad x_0 = \frac{\pi}{4}, \quad v_0 = 0, \quad \Delta t = 0.01$

(d) $w = 5, \quad m = 1, \quad x_0 = \frac{\pi}{2}, \quad v_0 = 0, \quad \Delta t = 0.01$

(e) $w = 5, \quad m = 5, \quad x_0 = \frac{\pi}{2}, \quad v_0 = 0, \quad \Delta t = 0.01$

(f) $w = 1, \quad m = 1, \quad x_0 = \frac{\pi}{2}, \quad v_0 = 1, \quad \Delta t = 0.01$

(g) $w = 1, \quad m = 1, \quad x_0 = \frac{\pi}{2}, \quad v_0 = -1, \quad \Delta t = 0.01.$

.5 Generate the positions only, using (4.31), of the harmonic oscillator with the parameters $w = 1$, $m = 1$, $x_0 = \frac{\pi}{4}$, $v_0 = 0$, $\Delta t = 0.01$. How do these results compare with those of Exercise 4.4(c)?

.6 For $w = m = 1$, does there exist a formula like (4.31) for v_{k+2}, v_{k+1} and v_k from which one could generate only v values from v_0 and v_1? Prove your answer.

.7 Consider a van der Pol oscillator with a_k defined by (4.46). Describe the motion of the oscillator for each of the following sets of parameter values. Which, if any, is periodic?

(a) $\lambda = 0.1, \quad m = 1, \quad x_0 = 2.500, \quad v_0 = 0, \quad \Delta t = 0.001$

(b) $\lambda = 0.1, \quad m = 1, \quad x_0 = 2.000, \quad v_0 = 0, \quad \Delta t = 0.001$

(c) $\lambda = 0.1, \quad m = 1, \quad x_0 = 1.500, \quad v_0 = 0, \quad \Delta t = 0.001$

(d) $\lambda = 1.0, \quad m = 1, \quad x_0 = 2.009, \quad v_0 = 0, \quad \Delta t = 0.001$

(e) $\lambda = 10, \quad m = 1, \quad x_0 = 3.006, \quad v_0 = 0, \quad \Delta t = 0.001$

(f) $\lambda = 10, \quad m = 1, \quad x_0 = 1.007, \quad v_0 = 0, \quad \Delta t = 0.001.$

.8 Show that, for the harmonic oscillator,

$$F_k = -\frac{V_{k+1}-V_k}{x_{k+1}-x_k}, \quad x_{k+1} \neq x_k,$$

where

$$V_k = \frac{w}{2} x_k^2.$$

CHAPTER 5

Waves

5.1 INTRODUCTION

Physical scientists are interested, for example, in light waves, sound waves, wat waves, electromagnetic waves, shock waves, and gravity waves, which should indica the depth to which the wave concept underlies the study of theoretical models. I this chapter we will study waves which are generated by a vibrating, elastic strin The reason for this choice is that a vibrating string can be understood easily an can be modeled by means of the oscillator concepts developed in the last chapter.

5.2 THE DISCRETE STRING

A discrete string is one which is composed of a finite number of particles in the following way. Let $x_k = k\Delta x$, $k = 0,1,2,\ldots,n+1$ be an R_{n+2} set. Then our particles are the ordered set P_k, $k = 0,1,2,\ldots,n,n+1$, with respective center (x_k,y_k), as shown typically in Fig. 5.1.

The prototype problem to be considered will be that of describing the return of a discrete string to a position of equilibrium from an initial position of tension. The resulting motion is considered to be an approximation to that of a real strin the improvement of which depends largely on one's computer capability to include more particles and more complex configurations.

We will assume that P_0 and P_{n+1} are fixed while $P_1,P_2,\ldots,P_n$ are free to mo <u>vertically only</u>, thus reducing the problem to the analysis of a system of oscilla tors. The assumption that horizontal motions are negligible is a popular one and seems to agree with many experimental results. However, we make it only for simplicity at the present time, since we can, in fact, easily incoporate horizontal motion into our model. The string motions which result are called transverse vibrations.

Throughout, let $x_0 = y_0 = y_{n+1} = 0$. For $\Delta x > 0$, let $x_j = j(\Delta x)$, $j = 0,1,2,\ldots,n+1$, be an R_{n+2} set. For $\Delta t > 0$, let $t_k = k\Delta t$, $k = 0,1,2,\ldots$, be measured in seconds, and let P_j, as shown in Fig. 5.2, be a typical particle in

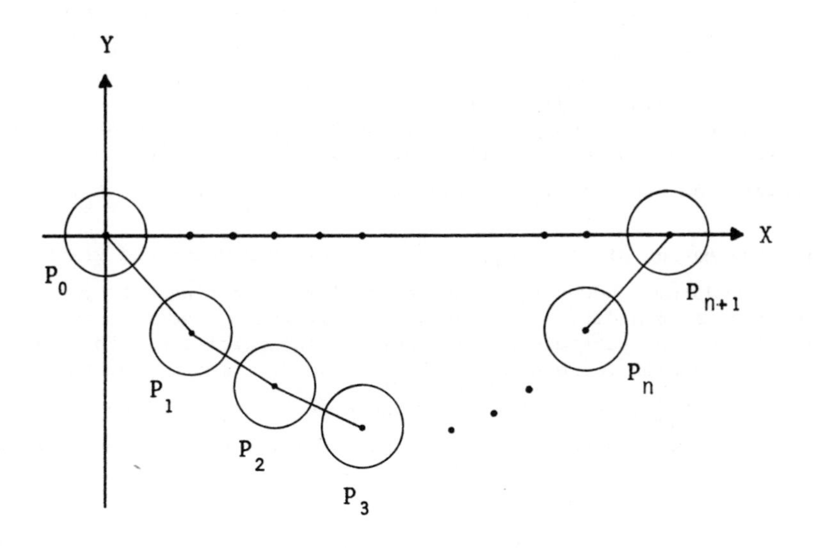

FIGURE 5.1

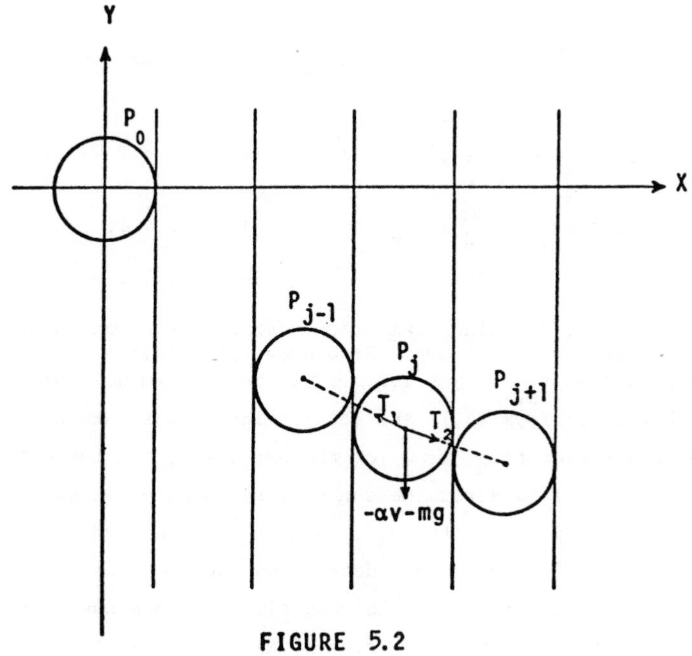

FIGURE 5.2

motion. In order to incorporate the time-dependence of the centers of P_{j-1}, P_j and P_{j+1}, let the respective centers of these particles at time t_k be $(x_{j-1},y_{j-1,k})$, $(x_j,y_{j,k})$, $(x_{j+1},y_{j+1,k})$, where each coordinate is measured in feet. (We use these units only for consistency with previous discussions and because we wish to allow for the effect of gravity). Also, let $a_{j,k}$ and $v_{j,k}$ be the acceleration and velocity, respectively, of P_j at time t_k.

In studying the motion of each P_j, let us consider damping and gravity, as in o study of the pendulum. However, if our particles are not each to go their separat ways, we must have an additional force which holds them together, that is, a force of tension between any particle and its neighboring particles. For this purpose, let T_1 be the tensile force between P_{j-1} and P_j, let T_2 be the tensile forc between P_j and P_{j+1}, and let the damping effect vary with the velocity of the particle. Then, for each particle P_j, the total force $F_{j,k}$ at time t_k is taken to be

$$F_{j,k} = |T_2| \frac{y_{j+1,k}-y_{j,k}}{[(\Delta x)^2+(y_{j+1,k}-y_{j,k})^2]^{1/2}} - |T_1| \frac{y_{j,k}-y_{j-1,k}}{[(\Delta x)^2+(y_{j,k}-y_{j-1,k})^2]^{1/2}}$$

$$- \alpha v_{j,k} - 32m; \quad j = 1,2,\dots,n, \quad k = 0,1,2,\dots \tag{5.1}$$

where $\alpha \geq 0$ and m is the mass of each particle. Also, we have

$$F_{j,k} = ma_{j,k}. \tag{5.2}$$

Since $v_{j,k}$ and $a_{j,k}$ are in the Y direction only, we have, in analogy with (4.3') and (4.4'),

$$v_{j,k+1} = v_{j,k} + a_{j,k}(\Delta t); \quad j = 1,2,\dots,n, \quad k = 0,1,2,\dots \tag{5.3}$$

$$y_{j,k+1} = y_{j,k} + \frac{\Delta t}{2}(v_{j,k+1}+v_{j,k}); \quad j = 1,2,\dots,n, \quad k = 0,1,2,\dots \tag{5.4}$$

The precise steps, or, the <u>algorithm</u>, for generating the motion of a discrete string can be given now as follows. From prescribed initial particle positions $y_{j,0}$ and velocities $v_{j,0}$, $j = 1,2,\dots,n$, one determines initial accelerations $a_{j,0}$, $j = 1,2,\dots,n$, from (5.1) and (5.2). The results are used in (5.3) to determine the next velocities $v_{j,1}$ of the particles, and these velocities, in turn, are used in (5.4) to determine the corresponding positions $y_{j,1}$. To procee to the second time step, the data $y_{j,1}$ and $v_{j,1}$ are used in (5.1) - (5.2) to determine the $a_{j,1}$, from which one determines the $v_{j,2}$ by means of (5.3), and then the $y_{j,2}$ by means of (5.4). In general, one steps ahead in time from $y_{j,k}$ and $v_{j,k}$ by determining the $a_{j,k}$ from (5.1) and (5.2), then the $v_{j,k+1}$ from (5.3), and finally the $y_{j,k+1}$ from (5.4). The computational procedure then is ompletely analogous to that given for the pendulum except that this time we have o determine the motions of n particles, rather than one particle, at each time tep.

5.3 EXAMPLES

et us consider now some simple string examples. In constructing these, the most ifficult problem is to decide upon the formulas for T_1 and T_2 in (5.1).

e most common practice in choosing T_1 and T_2 is to assume that each varies irectly with the distance between the respective particles involved. Thus, for xample, if the distance between P_{j-1} and P_j in Fig. 5.2 is r_1, then one ould assume

$$|T_1| = cr_1, \quad c > 0, \tag{5.5}$$

ere c is a constant which depends only on the physical nature of the string. rmula (5.5) is a particular example of what is called Hooke's law.

ysical experimentation, however, reveals that Hooke's law is only of limited lue. Indeed, (5.5) implies that if T_1 is arbitrarily large, so is r_1, and is simply is not correct. What happens, in fact, is that as T_1 continues to crease, r_1 increases less and less with T_1, until an elastic limit is reached d then the string actually breaks. Such qualitative behavior is called nonlinear, ile (5.5) is called linear.

our computer examples, then, we will try to simulate fully nonlinear behavior by ggesting and then using various possible formulas for T_1 and T_2 which may, or y not, include Hooke's law as a special case.

xample 1. Consider a twenty one particle string with $x = 0.1$; $x_j = \frac{j}{10}$, $j = 0,$ $2,\ldots,20$: $\alpha = 0.15$, $m = 0.05$, and $n = 19$. This string is relatively heavy d, however we start its motion, it should end up in a "steady state" position ich is not horizontal. In (5.1), let us set

$$T_1 = T_0\left[1 + \left|\frac{y_{j,k}-y_{j-1,k}}{\Delta x}\right| + \frac{\varepsilon}{2}\left|\frac{y_{j,k}-y_{j-1,k}}{\Delta x}\right|^2\right] \tag{5.6}$$

$$T_2 = T_0\left[1 + \left|\frac{y_{j+1,k}-y_{j,k}}{\Delta x}\right| + \frac{\varepsilon}{2}\left|\frac{y_{j+1,k}-y_{j,k}}{\Delta x}\right|^2\right]. \tag{5.7}$$

rmulas (5.6) and (5.7) are simple nonlinear relationships which describe the ten- on between successive particles as a function of the slope of the segment joining e centers of these particles. The string is placed in a position of tension by inging the center particle to (1,1), while those to the left of the center are sitioned on $y = x$ and those to the right of the center are positioned on $= -x + 2$. The resulting configuration is that shown for $t = 0$ in Fig. 5.3. e string is released from this position of tension, so that the initial velocity each particle is zero. For $T_0 = 12.5$, $\varepsilon = 0.01$, and $\Delta t = 0.00025$, its wnward motion from $t = 0$ to $t = 0.35$ is shown typically in Fig. 5.3, while s upward motion from $t = 0.35$ to $t = 0.69$ is shown typically in Fig. 5.4.

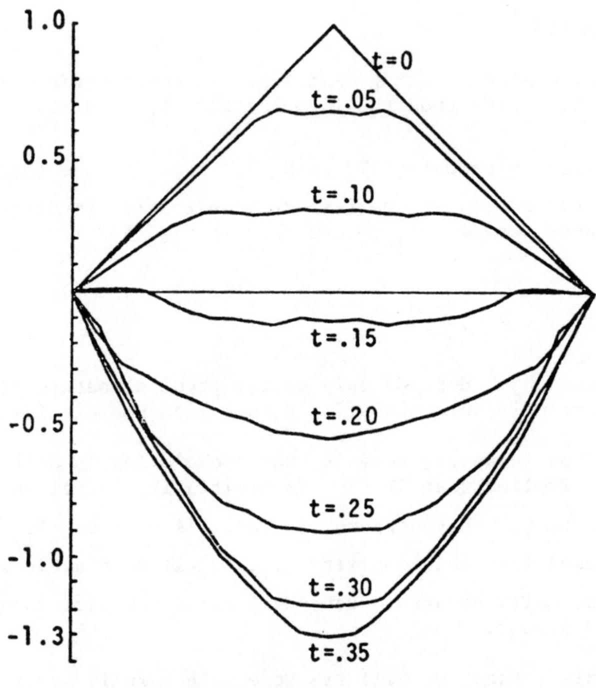

FIGURE 5.3

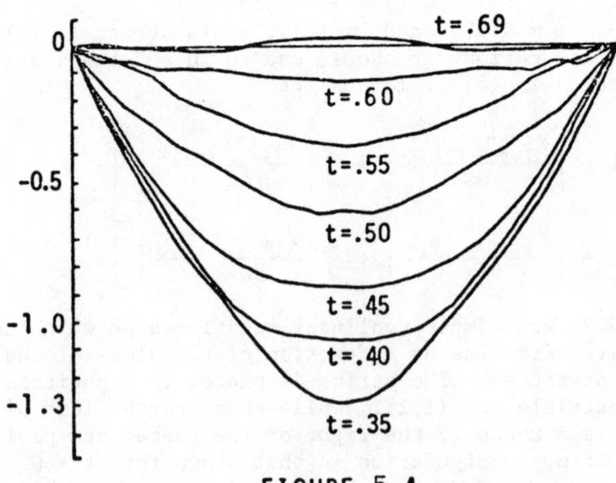

FIGURE 5.4

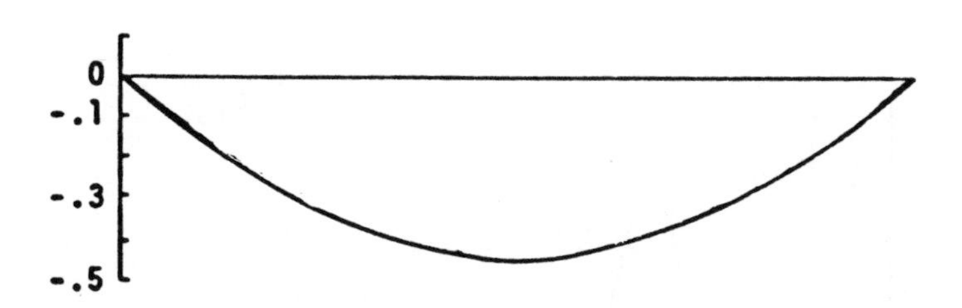

FIGURE 5.5

Fig. 5.5 is shown the string's position after six seconds, at which time it is
ose to steady state, with each particle vibrating a distance of at most 0.005.

xample 2. Consider again the string of Example 1, but with two changes. First,
et us neglect gravity by simply deleting the term $-32m$ from (5.1). Second, let
set the string in the following, different, initial position. The first moving
rticle is placed at (0.1, 0.5), the second at (0.2, 1.0), and the remaining
e centered on $y = -\frac{5}{9}(x-2)$, as shown at $t = 0.00$ in Fig. 5.6. The first 0.75
conds of motion is shown typically in Fig. 5.6. Convergence to a horizontal
eady state is at a rate comparable to that of Example 1. The actual reflection
the wave as it moves from left to right is seen from $t = .40$ to $t = .70$. A

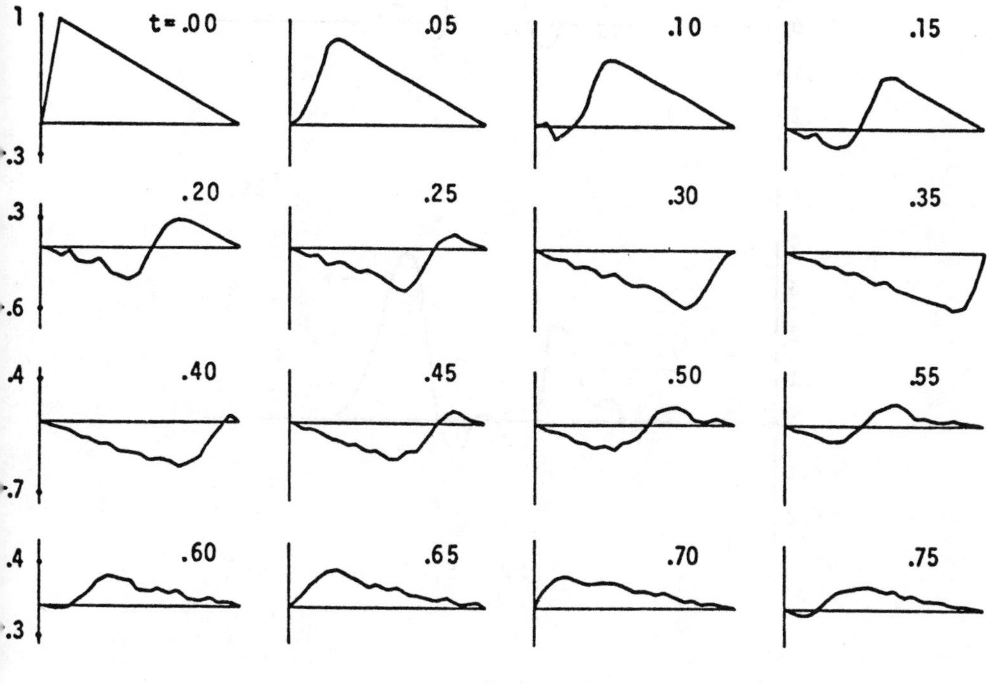

FIGURE 5.6

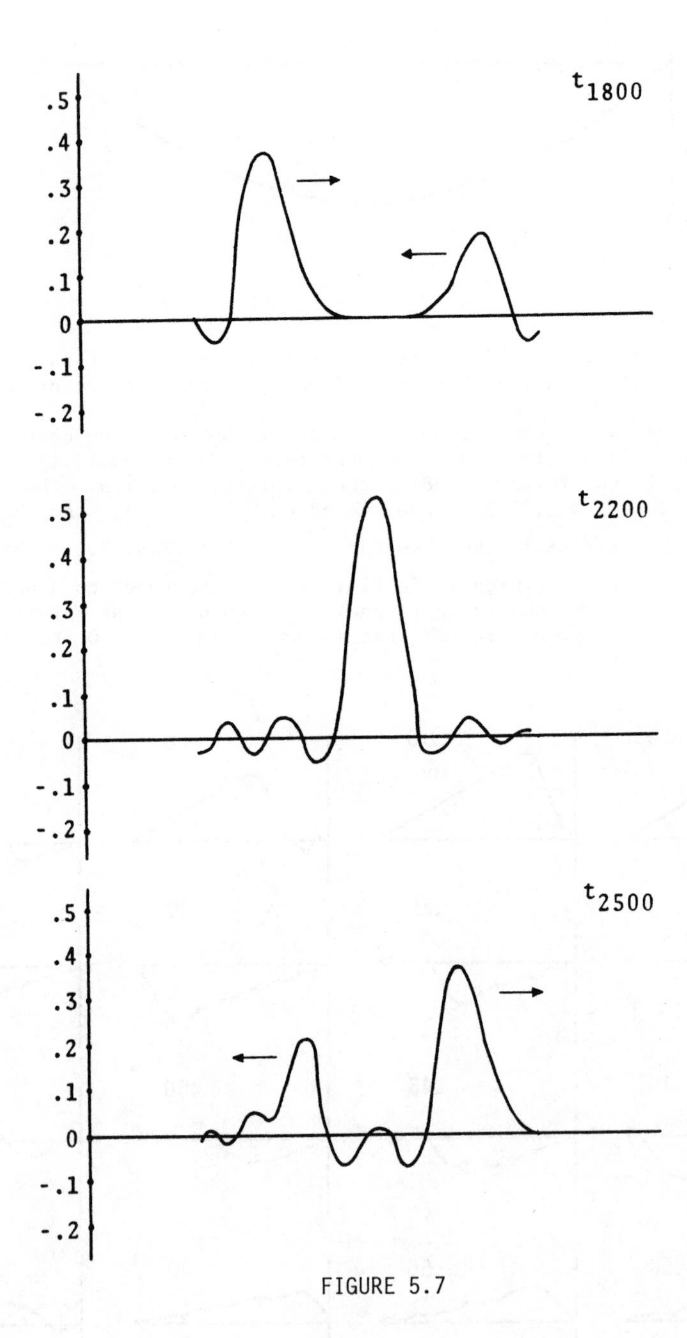

FIGURE 5.7

ery interesting aspect of the motion is seen in all the figures after $t = .15$, hat is, there are developing smaller waves which follow the main wave and these re called trailing waves.

xample 3. In our final example, we will not only change the tension formula, but ill also study the interesting phenomenon of two waves which collide.

irst, in (5.1), let us set

$$T_1 = T_0\left\{(1-\varepsilon)\left[\frac{\sqrt{(\Delta x)^2+(y_{j,k}-y_{j-1,k})^2}}{\Delta x}\right] + \varepsilon\left[\frac{\sqrt{(\Delta x)^2+(y_{j,k}-y_{j-1,k})^2}}{\Delta x}\right]^2\right\} \tag{5.8}$$

$$T_2 = T_0\left\{(1-\varepsilon)\left[\frac{\sqrt{(\Delta x)^2+(y_{j+1,k}-y_{j,k})^2}}{\Delta x}\right] + \varepsilon\left[\frac{\sqrt{(\Delta x)^2+(y_{j+1,k}-y_{j,k})^2}}{\Delta x}\right]^2\right\}, \tag{5.9}$$

here $0 \leq \varepsilon \leq 1$. In the special case where $\varepsilon = 0$, (5.8) and (5.9) reduce to ooke's law (5.5) with $c = T_0/\Delta x$. Next, since we will be concerned with the nteraction of two waves, which is an interaction of relatively short duration, let s neglect the effect of gravity by deleting the term $-32m$ from (5.1). Finally, n order to generate two waves which collide, at t_0 set all the particles on the -axis and give only particles at both ends non-zero initial velocities. In this ashion we will generate two waves, one at either end, which will move towards each ther and collide. The particular set of parameters which we choose are $T_0 = 10$, $= 0.01$, $\Delta t = 0.0001$, $\Delta x = 0.02$, $n = 99$, $m = .01$, $\varepsilon = 0.02$, $v_{1,0} = 60$, $_{2,0} = 50$, $v_{3,0} = 40$, $v_{4,0} = 30$, $v_{5,0} = 20$, $v_{6,0} = 10$, $v_{7,0} = v_{8,0} = \ldots =$ $_{95,0} = 0$, $v_{96,0} = 10$, $v_{97,0} = 20$, $v_{98,0} = 30$, $v_{99,0} = 40$. In Fig. 5.7 are hown the results of the collision, which is, effectively, that the waves pass hrough each other. A priori, this result would have been difficult to predict ith any assurity.

he FORTRAN program for this last example is given in Appendix B.

5.4 EXERCISES - CHAPTER 5

.1 Design and execute a physical experiment to test how well Example 2 of Section 5.3 simulates the motion of a light, elastic string. Can you detect trailing waves?

.2 Consider Example 1 of Section 5.3. For each of the following changes in the given data, describe the resulting vibration.

(a) Change only α and consider each of $\alpha = 0.2, 0.1, 0.01, 0.001$.

(b) Change only m and consider each of $m = 0.2, 0.1, 0.01, 0.001$.

(c) Change only the initial position as follows. Set particles P_5 and P_{15} at $(0.5, -1)$ and $(1.5, 1)$, respectively. Reset the centers of the particles between P_0 and P_5 vertically so that their centers are on the line $y = -2x$, those between P_5 and P_{15} vertically so that their centers are on the line $y = 2x - 2$, and those between P_{15} and P_{20} vertically so that their centers are on $y = -2x + 4$.

(d) Change the initial position as in (c), above, and also neglect gravity.

5.3 Produce a computer example, analogous to that of Example 3 of Section 5.3, which shows wave interaction, but use only 41 particles.

5.4 Produce a computer example, analogous to that of Example 3 of Section 5.3, which shows wave interaction, but use 501 particles.

5.5 Discuss the possibility of presenting computer results of vibrating string motions in the form of a movie.

5.6 Model and study a discrete vibrating string which consists of <u>three</u> rows of particles (rather than <u>one</u> row as shown in Fig. 5.1).

5.7 Model and study a vibrating violin string.

5.8 Hooke's law is considered to be a good approximation for "small" vibrations of a string. Do you think that Hooke's law would be adequate for studying the vibrations of a <u>metal</u> guitar string? Give two other instances in which Hooke's law would be adequate.

CHAPTER 6

Vectors

6.1 INTRODUCTION

we turn next to the study of particle motion in more than one direction, it will most convenient to have a clear and concise way to express physical relation-ips. This way will be provided by the concept of a vector and the use of vector tation, which will be developed in this chapter. The study of particle motion in re than one dimension will begin properly in Chapter 7.

6.2 TWO-DIMENSIONAL VECTORS

two-dimensional vector is nothing more than a complex number in disguise. Before vealing the disguise, let us recall some basic concepts related to complex num-rs.

r x and y real and $i = \sqrt{-1}$, the number $z = x + iy$ is called a <u>complex</u> mber. The <u>real</u> part of z is x and the imaginary part is iy. For any real nstant c, the product cz is defined by

$$cz = cx + icy. \tag{6.1}$$

$z_1 = x_1 + iy_1$ and $z_2 = x_2 + iy_2$ are two complex numbers, then $z_1 = z_2$ if d only if $x_1 = x_2$ and $y_1 = y_2$, while addition and subtraction are given by

$$z_1 + z_2 = (x_1+x_2) + i(y_1+y_2) \tag{6.2}$$

$$z_1 - z_2 = (x_1-x_2) + i(y_1-y_2). \tag{6.3}$$

ometrically, we plot the number z in a complex plane by requiring that the per-ndicular distance from z to the real X-axis is y and that the perpendicular stance to the imaginary Y-axis is X, as shown in Fig. 6.1. The distance from to the origin O is called the absolute value of z and is given by

$$|z| = (x^2+y^2)^{1/2}.$$

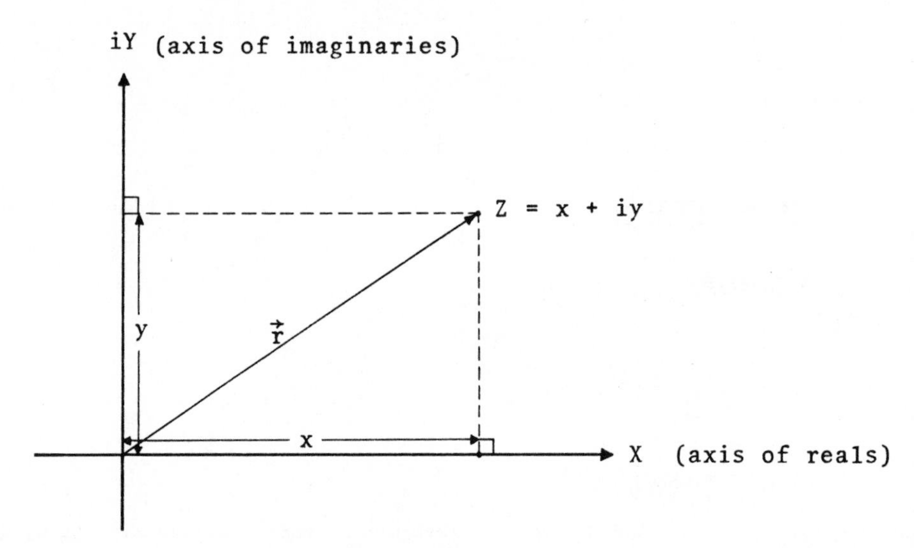

FIGURE 6.1

Now, to each number z in the complex plane, there corresponds a unique directed segment $\vec{r}$ from the origin to the point z. In addition, to each directed segment $\vec{r}$ from the origin of the complex plane, there corresponds a unique complex number z at the terminus of $\vec{r}$. With a little practice, both z and $\vec{r}$ can be used interchangeably. The quantity $\vec{r}$ is called a vector and is more useful in many physical situations than is the complex number z, and instead of using the notation $z = x + iy$, we eliminate the i by writing

$$\vec{r} = (x,y). \tag{6.4}$$

In (6.4), x and y are called the components of $\vec{r}$. Finally, we note that for vector in the form (6.4), one really doesn't need the use of the imaginary unit i so that in place of the iY-axis of Fig. 6.1, we can use the Y-axis itself. Thus, vectors given in the form (6.4) will be plotted in the real plane.

The fundamental concepts and rules for manipulation of complex numbers can now be translated into concepts and rules for vectors as follows. In analogy with (6.1), if $\vec{r} = (x,y)$ and c is a real constant, then

$$c\vec{r} = (cx,cy). \tag{6.5}$$

If $\vec{r}_1 = (x_1,y_1)$ and $\vec{r}_2 = (x_2,y_2)$, then $\vec{r}_1 = \vec{r}_2$ if and only if both $x_1 = x_2$ and $y_1 = y_2$, while, in analogy with (6.2) and (6.3), addition and subtraction of vectors are given by

$$\vec{r}_1 + \vec{r}_2 = (x_1+x_2,\ y_1+y_2) \tag{6.6}$$

$$\vec{r}_1 - \vec{r}_2 = (x_1-x_2,\ y_1-y_2). \tag{6.7}$$

The geometric meaning of (6.6) is especially interesting. For suppose, as shown in Fig. 6.2, one wishes to add the two vectors $\vec{r}_1$ and $\vec{r}_2$. Then the vector $\vec{r}_3$ is directed from the origin to the point $R(x_1+x_2,\ y_1+y_2)$. But, OPRQ is a

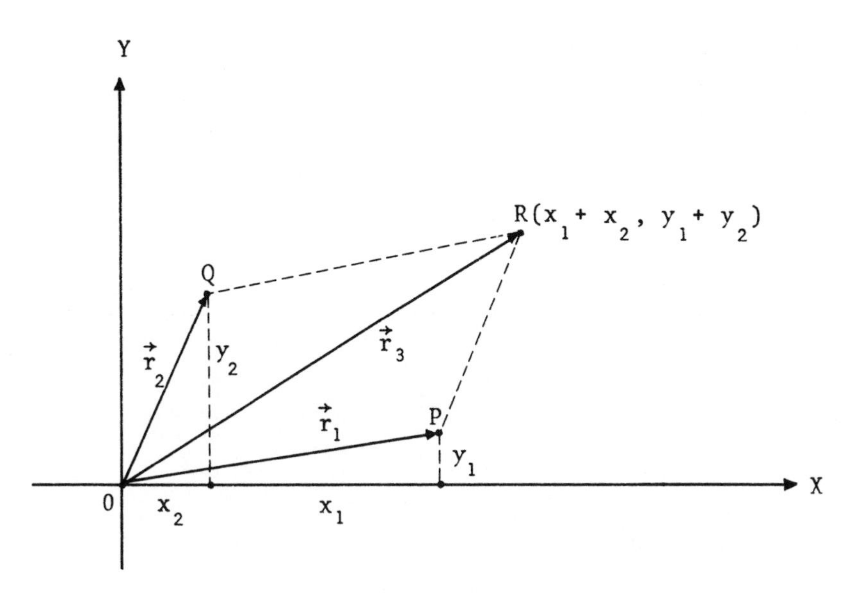

FIGURE 6.2

arallelogram. Thus, (6.6) is called the Law of Parallelograms for vector addiion.

6.3 THREE DIMENSIONAL VECTORS

t times it will be convenient to work in a plane, so that one can and should ecome adept with two dimensional vectors. But, as we shall see, there are pheomena we must analyze which are distinctly three dimensional. Thus, to proceed in n adequate fashion, it will be important to give now a comprehensive development f vectors in three dimensions. The concepts and results developed previously in wo dimensions will provide only the motivation for the way we will begin. The oncepts and results to be developed will, however, all be valid in two dimensions lso, since a two dimensional space is only a special portion of a three dimenional space.

o begin with, we will need a three dimensional, real XYZ coordinate system. his is usually drawn as in Fig. 6.3a or Fig. 6.3b. Both these systems are in comon use, with the Fig. 6.3a system being called a right-handed system, while the ther is called a left-handed system. For consistency, we will always work with a ight-handed system.

uppose now that one is given a point $P(x,y,z)$, as shown in Fig. 6.4. Then to ach such point there corresponds a unique directed segment $\vec{r}$ from the origin to . Conversely, to each such directed segment $\vec{r}$, there corresponds a unique point at the terminus. The quantity $\vec{r}$ is called a vector. It has both magnitude nd direction. Its magnitude $|\vec{r}| = r$ is the distance

$$|OP| = (x^2+y^2+z^2)^{1/2}.$$

any physical quantities of interest also have both magnitude and direction. Typial, for example, is velocity. When one expresses the velocity of a particle, one

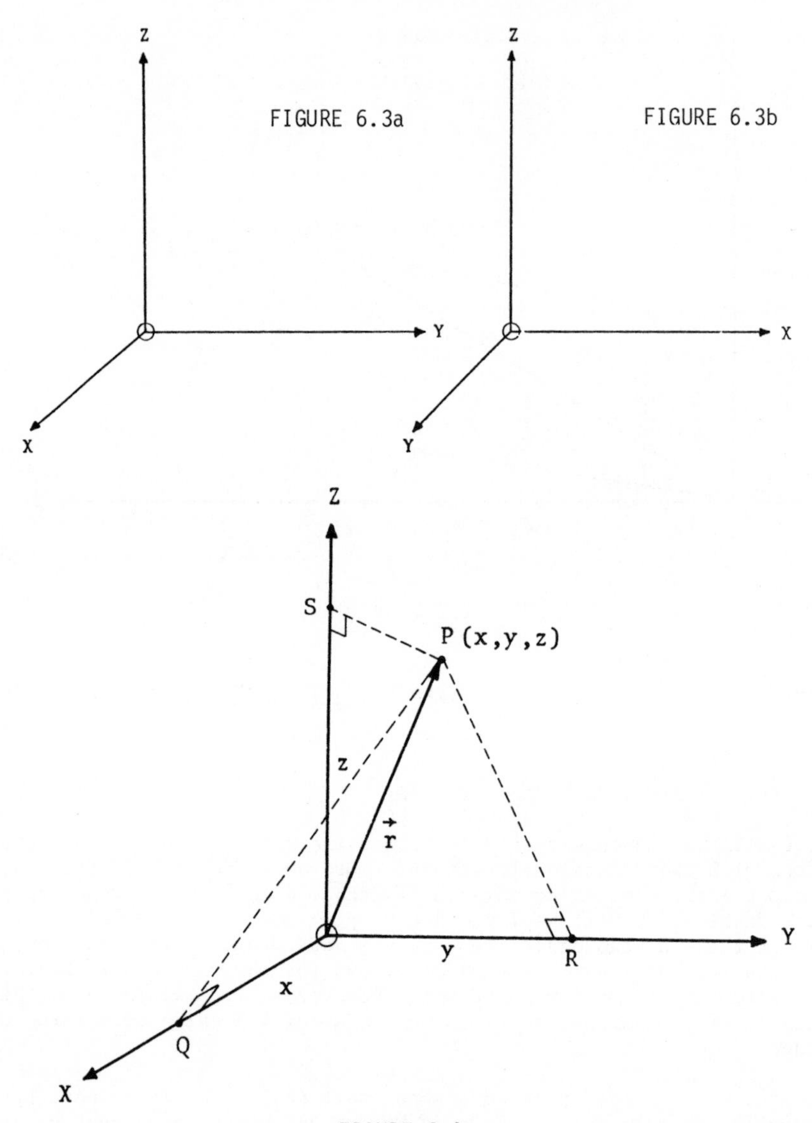

FIGURE 6.4

gives its direction and its speed. Thus, velocity can be represented by a vector whose <u>magnitude</u> is its speed. Speed is called a scalar, because no particular direction will be associated with it. It is speed plus a direction that yields velocity, which is a vector. We did not differentiate precisely between velocity and speed in one dimensional motion because the direction of a particle's motion is so limited in this case. However, in three dimensions, the differentiation between speed and velocity becomes essential. As we continue, other physical quantities, like <u>force</u> and <u>acceleration</u> will also be represented conveniently by vectors.

The coordinates of $P(x,y,z)$ are called the components of $\vec{r}$ and one writes

$$\vec{r} = (x,y,z).$$

hese coordinates express, intuitively, how much of $\vec{r}$ acts in the X, Y, and Z irections, respectively. With regard to the arithmetic of vectors, we assume, in nalogy with (6.5)-(6.7) that, for any real constant c,

$$c\vec{r} = (cx, cy, cz), \tag{6.8}$$

hile for any two vectors $\vec{r}_1 = (x_1, y_1, z_1)$, $\vec{r}_2 = (x_2, y_2, z_2)$, then $\vec{r}_1 = \vec{r}_2$ if nd only if $x_1 = x_2$, $y_1 = y_2$, $z_1 = z_2$, and that

$$\vec{r}_1 + \vec{r}_2 = (x_1+x_2,\ y_1+y_2,\ z_1+z_2) \tag{6.9}$$

$$\vec{r}_1 - \vec{r}_2 = (x_1-x_2,\ y_1-y_2,\ z_1-z_2). \tag{6.10}$$

e turn finally to the question of how vectors can be multiplied. For considera-ion of sums of products, like those of (2.25), we will develop one kind of product, alled a dot or inner product. To determine the components of a vector which is erpendicular to a given pair of vectors, a problem which would not have arisen, in eneral, in two dimensions, we will need a second kind of product called a cross roduct. Each of these will be developed in turn.

iven two vectors $\vec{r}_1 = (x_1, y_1, z_1)$, and $\vec{r}_2 = (x_2, y_2, z_2)$, the inner product $\vec{r}_1 \cdot \vec{r}_2$ is a scalar which is defined by

$$\vec{r}_1 \cdot \vec{r}_2 = x_1x_2 + y_1y_2 + z_1z_2. \tag{6.11}$$

he inner product is intimately related to the angle θ between two nonzero vec-ors $\vec{r}_1$ and $\vec{r}_2$. The exact relationship can be derived as follows. As shown in ig. 6.5, the law of cosines implies

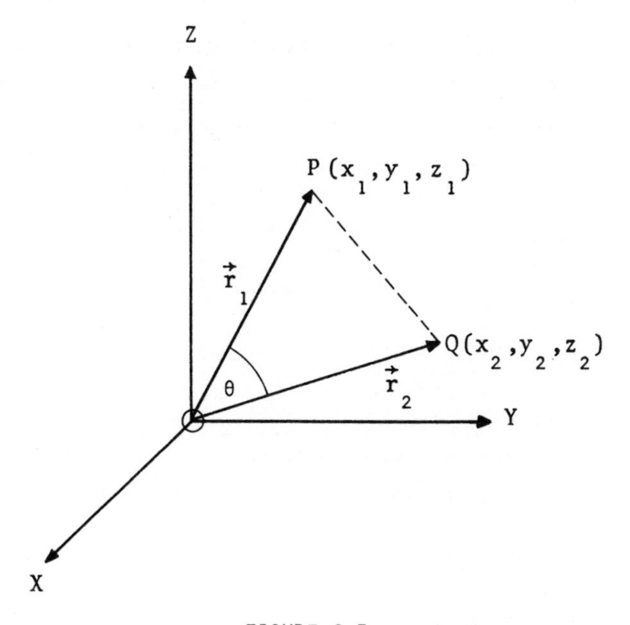

FIGURE 6.5

$$|PQ|^2 = r_1^2 + r_2^2 - 2r_1r_2 \cos\theta. \tag{6.12}$$

But,

$$r_1 = \sqrt{x_1^2+y_1^2+z_1^2}$$

$$r_2 = \sqrt{x_2^2+y_2^2+z_2^2}$$

$$|PQ| = \sqrt{(x_2-x_1)^2+(y_2-y_1)^2+(z_2-z_1)^2}\ ,$$

so that (6.12) yields easily

$$-2x_1x_2 - 2y_1y_2 - 2z_1z_2 = -2\sqrt{x_1^2+y_1^2+z_1^2}\ \sqrt{x_2^2+y_2^2+z_2^2}\ \cos\theta,$$

or, more simply,

$$\cos\theta = \frac{\vec{r}_1\cdot\vec{r}_2}{r_1r_2}. \tag{6.13}$$

From (6.13) follows the useful result that $\vec{r}_1$ is perpendicular to $\vec{r}_2$ if $\vec{r}_1 \cdot \vec{r}_2 = 0$. Note also that (6.11) implies both

$$\vec{r}_1 \cdot \vec{r}_2 = \vec{r}_2 \cdot \vec{r}_1 \tag{6.14}$$

$$\vec{r}_1 \cdot (\vec{r}_2+\vec{r}_3) = \vec{r}_1 \cdot \vec{r}_2 + \vec{r}_1 \cdot \vec{r}_3. \tag{6.15}$$

Finally, let us consider the problem of finding a vector $\vec{r}_3$ which is perpendicular to two nonzero, noncollinear vectors $\vec{r}_1$ and $\vec{r}_2$. For this purpose, let $\vec{r}_1 = (x_1,y_1,z_1)$, $\vec{r}_2 = (x_2,y_2,z_2)$ and $\vec{r}_3 = (x_3,y_3,z_3)$. Then, since $\vec{r}_3$ is to be perpendicular to $\vec{r}_1$, one has

$$x_1x_3 + y_1y_3 + z_1z_3 = 0. \tag{6.16}$$

Similarly, since $\vec{r}_3$ is to be perpendicular to $\vec{r}_2$, one has

$$x_2x_3 + y_2y_3 + z_2z_3 = 0. \tag{6.17}$$

In system (6.16) and (6.17), x_1, y_1, z_1, x_2, y_2 and z_2 are assumed to be known, so that we have to determine the three unknowns x_3, y_3 and z_3 from only two equations. To do this, let z_3 be arbitrary at present and rewrite (6.16) - (6.17)

$$x_1x_3 + y_1y_3 = -z_1z_3 \tag{6.18}$$

$$x_2x_3 + y_2y_3 = -z_2z_3. \tag{6.19}$$

olving this system for x_3 and y_3 yields

$$x_3 = \frac{\begin{vmatrix} -z_1z_3 & y_1 \\ -z_2z_3 & y_2 \end{vmatrix}}{\begin{vmatrix} x_1 & y_1 \\ x_2 & y_2 \end{vmatrix}} = z_3 \frac{\begin{vmatrix} -z_1 & y_1 \\ -z_2 & y_2 \end{vmatrix}}{\begin{vmatrix} x_1 & y_1 \\ x_2 & y_2 \end{vmatrix}} \tag{6.20}$$

$$y_3 = \frac{\begin{vmatrix} x_1 & -z_1z_3 \\ x_2 & -z_2z_3 \end{vmatrix}}{\begin{vmatrix} x_1 & y_1 \\ x_2 & y_2 \end{vmatrix}} = z_3 \frac{\begin{vmatrix} x_1 & -z_1 \\ x_2 & -z_2 \end{vmatrix}}{\begin{vmatrix} x_1 & y_1 \\ x_2 & y_2 \end{vmatrix}}. \tag{6.21}$$

e now choose z_3 by

$$z_3 = \begin{vmatrix} x_1 & y_1 \\ x_2 & y_2 \end{vmatrix}, \tag{6.22}$$

o that x_3 and y_3 in (6.20) and (6.21) reduce to relatively simple forms:

$$x_3 = \begin{vmatrix} -z_1 & y_1 \\ -z_2 & y_2 \end{vmatrix} \tag{6.23}$$

$$y_3 = \begin{vmatrix} x_1 & -z_1 \\ x_2 & -z_2 \end{vmatrix}. \tag{6.24}$$

hus, our choice for $\vec{r}_3$ has the components

$$x_3 = \begin{vmatrix} y_1 & z_1 \\ y_2 & z_2 \end{vmatrix}, \quad y_3 = \begin{vmatrix} z_1 & x_1 \\ z_2 & x_2 \end{vmatrix}, \quad z_3 = \begin{vmatrix} x_1 & y_1 \\ x_2 & y_2 \end{vmatrix} \tag{6.25}$$

nd it is called the <u>cross product</u> vector of $\vec{r}_1$ and $\vec{r}_2$, written more concisely s

$$\vec{r}_3 = \vec{r}_1 \times \vec{r}_2.$$

n the special case where $\vec{r}_1$ and $\vec{r}_2$ are <u>collinear</u>, the vector $\vec{r}_3$ is not nique, so that we simply define $\vec{r}_1 \times \vec{r}_2 = \vec{0}$. Thus, for example, for any vector one has $\vec{r} \times \vec{r} = \vec{0}$.

t follows readily from the definition of the cross product that

$$\vec{r}_1 \times \vec{r}_2 = -\vec{r}_2 \times \vec{r}_1 \tag{6.26}$$

$$\vec{r}_1 \times (\vec{r}_2+\vec{r}_3) = \vec{r}_1 \times \vec{r}_2 + \vec{r}_1 \times \vec{r}_3 \tag{6.27}$$

$$(C\vec{r}_1) \times \vec{r}_2 = C(\vec{r}_1\times\vec{r}_2), \quad C \text{ a constant} \tag{6.28}$$

$$\vec{r}_1 \cdot (\vec{r}_2 \times \vec{r}_3) = (\vec{r}_1 \times \vec{r}_2) \cdot \vec{r}_3 \quad (6.29)$$

$$\vec{r}_1 \times (\vec{r}_2 \times \vec{r}_3) = (\vec{r}_1 \times \vec{r}_2) \times \vec{r}_3 \quad (6.30)$$

$$\vec{r}_1 \times (\vec{r}_2 \times \vec{r}_3) = (\vec{r}_1 \cdot \vec{r}_3)\vec{r}_2 - (\vec{r}_1 \cdot \vec{r}_2)\vec{r}_3 \quad (6.31)$$

$$\vec{r}_1 \cdot (\vec{r}_1 \times \vec{r}_2) = \vec{0} \quad (6.32)$$

and

$$|\vec{r}_1 \times \vec{r}_2| = |\vec{r}_1| \cdot |\vec{r}_2| \sin \theta, \quad (6.33)$$

where θ is the angle between $\vec{r}_1$ and $\vec{r}_2$.

Note that (6.33) implies that increasing the magnitude of either $\vec{r}_1$ or $\vec{r}_2$ results in increasing the magnitude of the cross product vector $\vec{r}_1 \times \vec{r}_2$.

6.4 EXERCISES - CHAPTER 6

6.1 If $\vec{r}_1 = (5, 12)$, $\vec{r}_2 = (-1, 2)$, $\vec{r}_3 = (1, -3)$ are two dimensional vectors,

(a) draw each vector in the XY plane,

(b) determine the magnitude of each vector,

(c) determine each of $\vec{r}_1 + \vec{r}_2$, $\vec{r}_2 - \vec{r}_3$, $\vec{r}_1 + (\vec{r}_2 + \vec{r}_3)$, $\vec{r}_3 - (\vec{r}_1 - \vec{r}_2)$.

6.2 If $\vec{r}_1 = (0, 0)$, $\vec{r}_2 = (-3, 4)$, $\vec{r}_3 = (3, -4)$, $\vec{r}_4 = (1, -1)$ are two dimensional vectors,

(a) draw each vector in the XY plane,

(b) determine the magnitude of each vector,

(c) determine each of $\vec{r}_1 + 2\vec{r}_2$, $\vec{r}_3 - 4\vec{r}_4$, $(\vec{r}_1 + \vec{r}_2) - \vec{r}_3$, $\vec{r}_1 + (\vec{r}_2 - \vec{r}_3)$, $3\vec{r}_4 + (\vec{r}_2 - 2\vec{r}_3)$, $2\vec{r}_2 + 3(\vec{r}_4 - \vec{r}_3)$, $2(\vec{r}_1 + \vec{r}_2) - 3(\vec{r}_3 - 2\vec{r}_4)$.

6.3 Show that if $\vec{r}_1$ and $\vec{r}_2$ are nonzero, two dimensional vectors with the same direction, but with different magnitudes, then $\vec{r}_1 = c\vec{r}_2$, where c is a nonzero constant.

6.4 Show that if $\vec{r}_1$, $\vec{r}_2$ and $\vec{r}_3$ are two dimensional vectors, then

$$(\vec{r}_1 + \vec{r}_2) + \vec{r}_3 = \vec{r}_1 + (\vec{r}_2 + \vec{r}_3).$$

6.5 If $\vec{r}_1 = (2, 3, 4)$, $\vec{r}_2 = (-2, 3, 4)$, $\vec{r}_3 = (2, -3, 4)$ are three dimensional vectors,

(a) draw each vector in XYZ space,

(b) determine the magnitude of each vector,

(c) determine each of $\vec{r}_1 + \vec{r}_2$, $\vec{r}_3 - \vec{r}_2$, $\vec{r}_1 + (\vec{r}_2+\vec{r}_3)$, $\vec{r}_3 - (\vec{r}_1-\vec{r}_2)$.

6.6 If $\vec{r}_1 = (0, 0, 0)$, $\vec{r}_2 = (1, 1, 0)$, $\vec{r}_3 = (2, 3, 4)$, $\vec{r}_4 = (-2, -3, -4)$,

(a) draw each vector in XYZ space

(b) determine the magnitude of each vector,

(c) determine each of $\vec{r}_1 + 2\vec{r}_2$, $2\vec{r}_3 - 3\vec{r}_4$, $\vec{r}_2 + 3(\vec{r}_3-2\vec{r}_4)$, $(\vec{r}_1+\vec{r}_2) - (\vec{r}_3+\vec{r}_4)$, $(2\vec{r}_1+3\vec{r}_2) - (2\vec{r}_3+3\vec{r}_4)$, $2(\vec{r}_1-3\vec{r}_2) - 4(3\vec{r}_4-\vec{r}_3)$.

6.7 Show that if $\vec{r}_1$ and $\vec{r}_2$ are nonzero, three dimensional vectors with the same direction, but with different magnitudes, then $\vec{r}_1 = c\vec{r}_2$, where c is a nonzero constant.

6.8 Show that if $\vec{r}_1$, $\vec{r}_2$ and $\vec{r}_3$ are three dimensional vectors, then $(\vec{r}_1+\vec{r}_2)+\vec{r}_3 = \vec{r}_1 + (\vec{r}_2+\vec{r}_3)$.

6.9 For the vectors given in Exercise 6.5, determine $\vec{r}_1 \cdot \vec{r}_2$, $\vec{r}_2 \cdot \vec{r}_1$, $\vec{r}_1 \cdot \vec{r}_3$, $\vec{r}_2 \cdot \vec{r}_3$.

6.10 For the vectors given in Exercise 6.6, determine $\vec{r}_1 \cdot \vec{r}_2$, $\vec{r}_1 \cdot \vec{r}_4$, $\vec{r}_2 \cdot \vec{r}_3$, and $\vec{r}_3 \cdot \vec{r}_2$.

6.11 To the nearest degree, find the angle between each pair of vectors of Exercise 6.5.

6.12 To the nearest degree, find the angle between each pair of nonzero vectors of Exercise 6.6.

6.13 Which of the following pairs of vectors are perpendicular:

(a) (1, 0, 1), (-1, 2, 1)

(b) (2, 3, 4), (-2, -3, -4)

(c) (7, 9, 1), (-1, 9, -7)

(d) (2, 3, 4), (-3, 2, 0)

(e) (6, -6, 12), (-1, 1, 1).

(f) (a, b, -b), (b, a, 2a).

6.14 Show that the vectors of Exercise 6.5 satisfy (6.15).

6.15 Prove (6.15).

6.16 For each of the following, find $\vec{r}_1 \times \vec{r}_2$:

(a) $\vec{r}_1 = (1, 2, 3)$, $\vec{r}_2 = (3, 2, 1)$

(b) $\vec{r}_1 = (0, 1, 0)$, $\vec{r}_2 = (1, 0, 1)$

(c) $\vec{r}_1 = (-2, 0, 2)$, $\vec{r}_2 = (1, 1, 1)$

(d) $\vec{r}_1 = (-4, 7, 13)$, $\vec{r}_2 = (3, 3, 5)$

(e) $\vec{r}_1 = (1, 0, 0)$, $\vec{r}_2 = (0, 1, 0)$

(f) $\vec{r}_1 = (-1, -2, 1)$, $\vec{r}_2 = (3, -6, 7)$.

6.17 Show that each pair of vectors in Exercise 6.16 satisfies (6.26).

6.18 Prove (6.26).

6.19 Show that the vectors of Exercise 6.5 satisfy (6.27).

6.20 Prove (6.27).

6.21 Show that the vectors of Exercise 6.5 satisfy (6.29).

6.22 Prove (6.29).

6.23 Show that the vectors of Exercise 6.5 satisfy (6.30).

6.24 Prove (6.30).

6.25 Show that the vectors of Exercise 6.5 satisfy (6.31).

6.26 Prove (6.31).

6.27 Show that the vectors of Exercise 6.16(a) and 6.16(b) satisfy (6.32).

6.28 Prove (6.32).

6.29 Show that the vectors of Exercise 6.16(a) and 6.16(b) satisfy (6.33).

6.30 Prove (6.33).

CHAPTER 7

Gravitation

7.1 INTRODUCTION

ı every day usage, the terms gravity and gravitation often are confused and used ıterchangeably. We have, now, a fairly good understanding of gravity, so let us ırn next to the study of gravitation. In so doing, we will develop, simultane-ısly, the basic concepts and formulas for the motion of a particle in more than ıe dimension.

ıe term gravitation is used, generally, to describe the attraction between the ɔdies of our solar system. Since experiments related to heavenly bodies are dif-icult for us to carry out, in terms of both the equipment and the time required, ɜ must turn to the observations of others for help. Fortunately, man's interest ı the heavens has been almost universal, so that observational results are avail-ble from the times of antiquity. The refinement of man's understanding of gravi-ation was aided enormously by the works of Nicholas Copernicus and Johannes Kepler, ıt the master touch, in the form of a simple mathematical formula for the attrac-ive force between <u>all</u> bodies, heavenly or not, was the result, once again, of the ɔrk of Isaac Newton.

ı this chapter, then, we will develop a theory of gravitation similar to that of ɜwton. However, the more astute reader should note that the theory to be developed ıs, indeed, been refined further by Einstein's General Theory of Relativity, but ıat the experimental verification of General Relativity is still somewhat incom-lete.

7.2 THE $1/r^2$ LAW

ıroughout the study of physics, one often encounters what is called a "$1/r^2$ law." ıis will, indeed, be the case in our study of gravitation. In the present section, ıen, let us simply develop some intuition about such laws and some appreciation of ɔw they arise. For this purpose we will consider a "simple" situation in which a ight shines through a window and onto a wall, as follows.

ıppose light emanates from position 0 in Fig. 7.1, shines through the window BCD and onto the wall which is parallel and behind ABCD. Let the projection of BCD on the wall be A'B'C'D', as shown in Fig. 7.1. Now, the real nature of ight is, at present, not clearly understood, so, for the sake of argument, let us

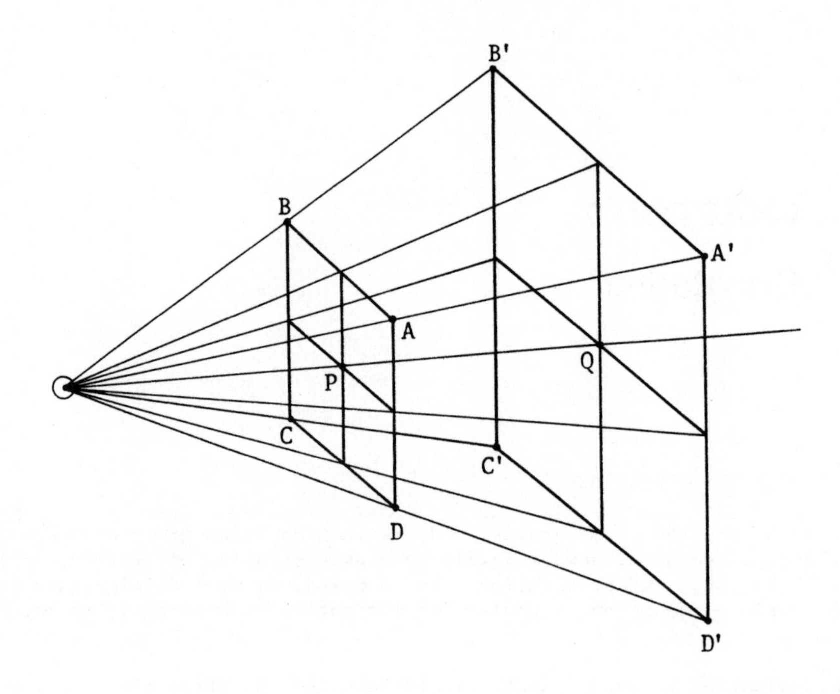

FIGURE 7.1

assume that light is emitted in little particles called quanta. Assume also that M quanta have reached ABCD at a given time and that these and only these quanta have reached A'B'C'D' at a later time. (This would be a reasonable assertion, for example, if the window and the wall are close to each other while O is relatively far away from both.) Then the light intensity I on ABCD is defined by

$$I = \frac{M}{\text{Area ABCD}}$$

while the light intensity I' on A'B'C'D' is defined by

$$I' = \frac{M}{\text{Area A'B'C'D'}}.$$

Hence,

$$I' = I\frac{\text{Area ABCD}}{\text{Area A'B'C'D'}}.$$

But,

$$\frac{\text{Area ABCD}}{\text{Area A'B'C'D'}} = \frac{(OP)^2}{(OQ)^2},$$

so that

$$I' = I\frac{(OP)^2}{(OQ)^2}.$$

If we now fix our units so that $I = 1$ and $OP = 1$, and if we set $OQ = r$, then

$$I' = \frac{1}{r^2},$$

or, the intensity I' varies inversely as the square of the distance from the source O. This behavior is called a "$1/r^2$ law" of variation. Such laws are convenient to assume and often enable one to construct a good, initial physical model. Our model of gravitation will be such a model, but with a constant of variation which is, in general, different from unity.

It is interesting to note that, historically, during the 1860's, J. C. Maxwell developed a unifying theory of electromagnetic waves, including light. This theory was verified experimentally and modified the $1/r^2$ law to include also a $1/r$ term, which becomes significant at large distances (see, e.g., Feynman, Leighton, and Sands).

7.3 GRAVITATION

Let us see now if we can develop some thinking about gravitation by beginning with gravity, as follows. Consider two bodies P_1 of mass m_1 and P_2 of mass m_2, each on an X-axis, as shown in Fig. 7.2. Assume at first that the mass of P_2 is almost negligible compared to that of P_1. Then, if P_1 were, for example, the earth, and P_2 were a particle near the earth, we would expect P_2 to fall to P_1 because of gravity. Now, suppose P_2 were a little more massive. It would probably still fall. Suppose it were even more massive. It might still fall. Suppose, indeed, that P_2 were as massive as P_1. Now it is not so clear as to what would happen. Moreover, if next we let P_2 be so massive that the mass of P_1 becomes the relatively negligible one, then, indeed, we would even expect P_1 to fall to P_2. This suggests the possibility that, in fact, P_1 and P_2 might both have been in motion in all cases, but that only when the mass of one particle was relatively negligible to the mass of the second did the motion of the second become relatively negligible with respect to the motion of the first. This behavior was assumed first by Isaac Newton and formalized in his Law of Gravitation, which is given as follows:

Newton's Law of Gravitation. Each of two bodies, P_1 of mass m_1 and P_2 of mass m_2, exerts an attractive force on the other which is called the force of gravitation. The force which P_1 exerts on P_2 is equal in magnitude but opposite in direction to the force which P_2 exerts on P_1, and both forces act along the straight line joining the centers of P_1 and P_2. The magnitude $|F|$ of each of these forces is given by the $1/r^2$ law:

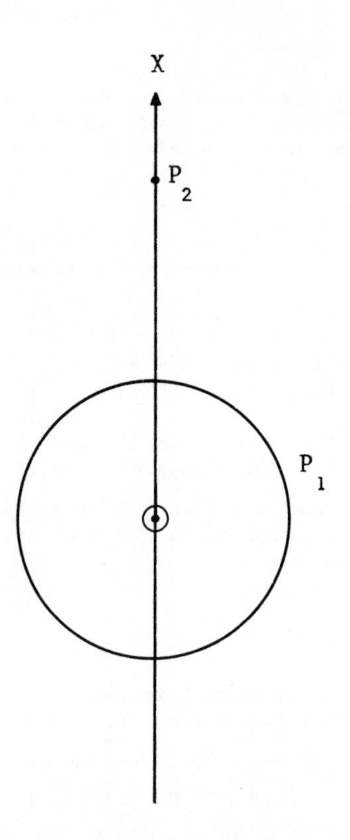

FIGURE 7.2

$$|F| = G\frac{m_1 m_2}{r^2}, \tag{7.1}$$

where r is the distance between the centers of P_1 and P_2 and where G is a universal constant, that is, G is a constant whose value does not depend in any way on P_1, P_2, or r.

Various experiments, which date back from those of H. Cavendish in 1798 to those currently being performed by various space agency programs, have determined in cgs (centimeter, gram, second) units the approximation

$$G = (6.67)10^{-8}. \tag{7.2}$$

With the intuition and ideas developed thus far, we are now ready to formulate a numerical model of Newtonian gravitation in more than one dimension. Of course, this is necessary if we wish to analyze, for example, the interesting motions of the planets. For this purpose, then, we must proceed first to extend the basic dynamical concepts of Chapter 3 to more than one dimension.

7.4 BASIC PLANAR CONCEPTS

Let us show first how to extend the basic concepts of position, velocity, and acceleration of Chapter 3. Since each of these quantities has both magnitude and direction, it is natural that we will introduce vector notation. But, interestingly enough, from the vector equations, themselves, which follow, one will not be able to deduce in how many dimensions one is modeling. Thus, the formulation, in essence, includes the three dimensional case, but for simplicity is developed as if one were interested at present only in the planar, or two dimensional, case.

For $\Delta t > 0$ and $t_k = k\Delta t$, $k = 0,1,2,\ldots$, let particle P of mass m be located at $\vec{r}_k = (x_k, y_k)$ at time t_k. If $\vec{v}_k = (v_{k,x}, v_{k,y})$ is the velocity of P at t_k, while $\vec{a}_k = (a_{k,x}, a_{k,y})$ is the acceleration of P at t_k, we will assume, in analogy with (3.5) and (3.6), that

$$\frac{\vec{v}_{k+1}+\vec{v}_k}{2} = \frac{\vec{r}_{k+1}-\vec{r}_k}{\Delta t}, \quad k = 0,1,2,\ldots, \tag{7.3}$$

$$\vec{a}_k = \frac{\vec{v}_{k+1}-\vec{v}_k}{\Delta t}, \quad k = 0,1,2,\ldots\ . \tag{7.4}$$

Of course, (7.3) and (7.4) are merely convenient and concise expressions for, respectively,

$$\frac{v_{k+1,x}+v_{k,x}}{2} = \frac{x_{k+1}-x_k}{\Delta t},$$

$$\frac{v_{k+1,y}+v_{k,y}}{2} = \frac{y_{k+1}-y_k}{\Delta t}, \quad k = 0,1,2,\ldots, \tag{7.3'}$$

$$a_{k,x} = \frac{v_{k+1,x}-v_{k,x}}{\Delta t},$$

$$a_{k,y} = \frac{v_{k+1,y}-v_{k,y}}{\Delta t}, \quad k = 0,1,2,\ldots\ . \tag{7.4'}$$

Thus, each vector equation in <u>two</u> dimensions represents <u>two</u> scalar equations.

Note that if one were in <u>three</u> dimensions, the single vector equation (7.4), for example, would be equivalent to the <u>three</u> scalar equations

$$a_{k,x} = \frac{v_{k+1,x}-v_{k,x}}{\Delta t}, \quad a_{k,y} = \frac{v_{k+1,y}-v_{k,y}}{\Delta t},$$

$$a_{k,z} = \frac{v_{k+1,z}-v_{k,z}}{\Delta t}, \quad k = 0,1,2,\ldots\ .$$

To relate force and acceleration at each time t_k, we assume, in analogy with (3.9), that

$$\vec{F}_k = m\vec{a}_k, \tag{7.5}$$

where

$$\vec{F}_k = (F_{k,x}, F_{k,y}). \tag{7.6}$$

Of course, (7.5) is equivalent to

$$F_{k,x} = ma_{k,x}, \quad F_{k,y} = ma_{k,y}, \quad k = 0,1,2,\ldots . \tag{7.5'}$$

In analogy with (3.12), next define the work W_n by

$$W_n = \sum_{k=0}^{n-1} [(\vec{r}_{k+1} - \vec{r}_k) \cdot \vec{F}_k]. \tag{7.7}$$

From the definition of the inner product, (7.7) is equivalent to

$$W_n = \sum_{k=0}^{n-1} [(x_{k+1} - x_k)F_{k,x} + (y_{k+1} - y_k)F_{k,y}]. \tag{7.8}$$

Now,

$$\sum_{k=0}^{n-1} (x_{k+1} - x_k)F_{k,x} = \sum_{0}^{n-1} (x_{k+1} - x_k)ma_{k,x} = m \sum_{0}^{n-1} [(\frac{x_{k+1} - x_k}{\Delta t})(v_{k+1,x} - v_{k,x})]$$

$$= \frac{m}{2} \sum_{0}^{n-1} [(v_{k+1,x} + v_{k,x})(v_{k+1,x} - v_{k,x})]$$

$$= \frac{m}{2} \sum_{0}^{n-1} (v_{k+1,x}^2 - v_{k,x}^2) = \frac{m}{2} v_{n,x}^2 - \frac{m}{2} v_{0,x}^2 .$$

Similarly,

$$\sum_{k=0}^{n-1} (y_{k+1} - y_k)F_{k,y} = \frac{m}{2} v_{n,y}^2 - \frac{m}{2} v_{0,y}^2 .$$

Thus, from (7.8),

$$W_n = \frac{m}{2}(v_{n,x}^2 + v_{n,y}^2) - \frac{m}{2}(v_{0,x}^2 + v_{0,y}^2). \tag{7.9}$$

We define now the kinetic energy K_i of P at time t_i by

$$K_i = \frac{m}{2}|\vec{v}_i|^2 = \frac{m}{2}(v_{i,x}^2 + v_{i,y}^2). \tag{7.10}$$

Then, from (7.9), one has

$$W_n = K_n - K_0, \tag{7.11}$$

in complete analogy with (3.13).

For convenience, we incorporate this into the following theorem, in analogy with Theorem 3.4.

heorem 7.1. For $\Delta t > 0$, let $t_k = k\Delta t$. Let particle P of mass m be in otion in an XY plane. At time t_k, let P be located at $\vec{r}_k$, have velocity k and have acceleration $\vec{a}_k$. If W_n is defined by (7.7), then (7.11) is valid or all $n = 1,2,3,\ldots$.

7.5 PLANETARY MOTION AND DISCRETE GRAVITATION

or simplicity, let us begin by studying a problem in which only one particle is in otion in the XY plane. A prototype problem of this type is the following, in hich a planet, whose mass is relatively small compared to that of the sun, is in rbital motion around the sun.

et the sun, whose mass is m_1, be positioned at the origin of the XY coordinate ystem. Let the position, velocity, and mass m_2 of a planet P be known at time 0. Then, assuming that the sun's motion is negligible, we must determine the osition (x_k,y_k) of P at each t_k, $k = 1,2,3,\ldots,n$, if the only acting force s gravitation. We define this gravitational force by the discrete formulas

$$F_k = (F_{k,x},F_{k,y}) \tag{7.12}$$

$$F_{k,x} = -\frac{Gm_1m_2}{r_kr_{k+1}}\,\frac{\frac{x_{k+1}+x_k}{2}}{\frac{r_{k+1}+r_k}{2}} = -\frac{Gm_1m_2(x_{k+1}+x_k)}{r_kr_{k+1}(r_k+r_{k+1})} \tag{7.13}$$

$$F_{k,y} = -\frac{Gm_1m_2(y_{k+1}+y_k)}{r_kr_{k+1}(r_k+r_{k+1})}, \tag{7.14}$$

here G is the Newtonian constant (7.2), and where

$$r_k^2 = x_k^2 + y_k^2, \tag{7.15}$$

ow, immediately, we must be concerned with the fact that (7.13) and (7.14) resemle (7.1) only vaguely, so this matter will have to be studied first.

ormulas (7.13) and (7.14) were developed, in fact, with hindsight, as follows. To egin with, there does not seem to be any reason to assume that damping is present n solar system motions to any significant degree. Hence, any theory of gravitaion should be energy conserving. Formulas (7.13) and (7.14) are energy conserving ormulas, as will be shown later, and this is why, after much pencil and paper work, hey were chosen. Let us show now, however, that they are indeed "$1/r^2$" formulas.

uppose, as shown in Fig. 7.3, that at time t_k, P is at (x_k,y_k). One would xpect the force $\vec{F}_k$ exerted by the sun on P to have the magnitude

$$|F| = \frac{Gm_1m_2}{r_k^2}. \tag{7.16}$$

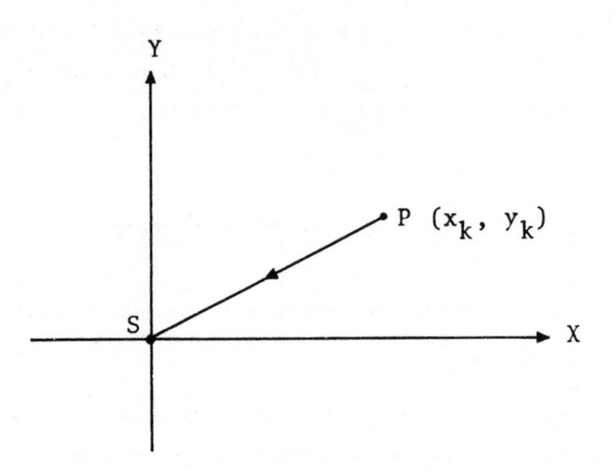

FIGURE 7.3

However, if (7.16) is the magnitude of $\vec{F}_k$, then the actual force is given by

$$\vec{F}_k = \left(-\frac{Gm_1m_2}{r_k^2}\frac{x_k}{r_k}, -\frac{Gm_1m_2}{r_k^2}\frac{y_k}{r_k}\right). \tag{7.17}$$

The components of (7.17) now appear to be analogous to (7.13) and (7.14). For example, comparing x-coefficients of (7.13) and (7.17), we have, approximately,

$$-\frac{Gm_1m_2}{r_k^2}\frac{x_k}{r_k} \sim -\frac{Gm_1m_2 \frac{x_{k+1}+x_k}{2}}{r_kr_{k+1}\frac{r_k+r_{k+1}}{2}},$$

that is, (7.13) uses the arithmetic mean $\frac{x_{k+1}+x_k}{2}$ for x_k, the arithmetic mean $\frac{r_{k+1}+r_k}{2}$ for r_k, and the product r_kr_{k+1} at successive time steps in place of r_k^2. If one could use a clock with arbitrarily small time steps, the differences between the above expressions would be negligible; so that, in effect, the $\frac{1}{r_kr_{k+}}$ product yields a $1/r^2$ law. It is of interest to note that the modern theory of quantum mechanics says that such a clock can be constructed, while the theory of relativity says that it cannot. Thus, the two modern theories of physics yield opposite conclusions on this point.

Let us show next that our equations (7.12) - (7.14) conserve energy, as stated previously. To do this will require a suitable definition of potential energy, which is arrived at as follows.

Consider, again, (7.7). Then (7.12) - (7.15) imply

$$W_n = \sum_{k=0}^{n-1}\left[(x_{k+1}-x_k)\left(-\frac{Gm_1m_2(x_{k+1}+x_k)}{r_kr_{k+1}(r_k+r_{k+1})}\right) + (y_{k+1}-y_k)\left(-\frac{Gm_1m_2(y_{k+1}+y_k)}{r_kr_{k+1}(r_k+r_{k+1})}\right)\right]$$

$$= -Gm_1m_2\sum_0^{n-1}\left[\frac{x_{k+1}^2-x_k^2+y_{k+1}^2-y_k^2}{r_kr_{k+1}(r_k+r_{k+1})}\right] = -Gm_1m_2\sum_0^{n-1}\left[\frac{r_{k+1}^2-r_k^2}{r_kr_{k+1}(r_k+r_{k+1})}\right]$$

$$= -Gm_1m_2\sum_0^{n-1}\left[\frac{r_{k+1}-r_k}{r_kr_{k+1}}\right] = -Gm_1m_2\sum_0^{n-1}\left[\frac{1}{r_k}-\frac{1}{r_{k+1}}\right] = \frac{-Gm_1m_2}{r_0}+\frac{Gm_1m_2}{r_n}\;.$$

f one defines the gravitational potential energy V_k at t_k by

$$V_k = -\frac{Gm_1m_2}{r_k}\;, \tag{7.18}$$

hen

$$W_n = -V_n + V_0. \tag{7.19}$$

ence, (7.11) and (7.19) imply, by the elimination of W_n, that

$$K_n + V_n = K_0 + V_0, \quad n = 1,2,3,\ldots \tag{7.20}$$

hich is the law of conservation of energy.

nce again, it is worth noting that a telescopic sum, which in this case is

$$\sum_0^{n-1}\left[\frac{1}{r_k}-\frac{1}{r_{k+1}}\right],$$

lays a key role in the determination of the potential function V_k.

e wish to turn next to an actual computational example in which an orbit is con-
tructed. However, as will be seen, we will have to be able to solve some alge-
raic equations which are not like the trivial ones usually studied in the class-
oom. For this reason, we will discuss first an iterative, computer technique,
alled Newton's method, which will enable us to solve our problem.

7.6 NEWTON'S METHOD OF ITERATION

or simplicity, suppose one has to find a root of the single equation

$$f(x) = 0. \tag{7.21}$$

n algebra, one usually studies very well behaved equations like

$$3x - 4 = 0, \tag{7.22}$$

nd one learns how to find the root with exceptional ease. In science, however,
quations more like

$$x - 10^{-x} = 0 \tag{7.23}$$

occur very often, and one's first attempts to solve it may be somewhat traumatic, for it is a very difficult equation to solve. Indeed, no one knows how to solve it <u>exactly</u>. So we do the next best thing, that is, we try to get a very good approximation to a root, and the method to be developed now will allow us to obtain such an approximation to within any predetermined accuracy. This method is an iterative method and is named after its developer, Isaac Newton.

Geometrically, the problem of finding a root of the equation (7.21) is equivalent to that of finding a zero of the function $y = f(x)$, that is, as shown in Fig. 7.4 we seek to find the point $\bar{x}$ where the graph of $y = f(x)$ crosses the X-axis, for $f(\bar{x}) = 0$. The general idea for doing this is as follows. Guess an approximation $x^{(0)}$ to $\bar{x}$. It is most probably incorrect, so we try to use it to get a better approximation $x^{(1)}$ to $\bar{x}$ in the following way. For $\Delta x > 0$, find the two points $(x^{(0)}, f(x^0))$ and $(x^{(0)}+\Delta x, f(x^{(0)}+\Delta x))$ on the curve $y = f(x)$, as shown in Fig 7.4. Draw the line through these two points, and where it crosses the X-axis, take this intercept to be $x^{(1)}$. The exact formula for $x^{(1)}$ is obtained in the following way. The equation of the straight line through $(x^{(0)}, f(x^{(0)}))$ and $(x^{(0)}+\Delta x, f(x^{(0)}+\Delta x))$ is

$$y - f(x^{(0)}) = \frac{f(x^{(0)}+\Delta x) - f(x^{(0)})}{\Delta x}(x - x^{(0)}). \tag{7.24}$$

The X-intercept $x^{(1)}$ of (7.24) is found by setting $y = 0$ in (7.24) and solving for x, so that

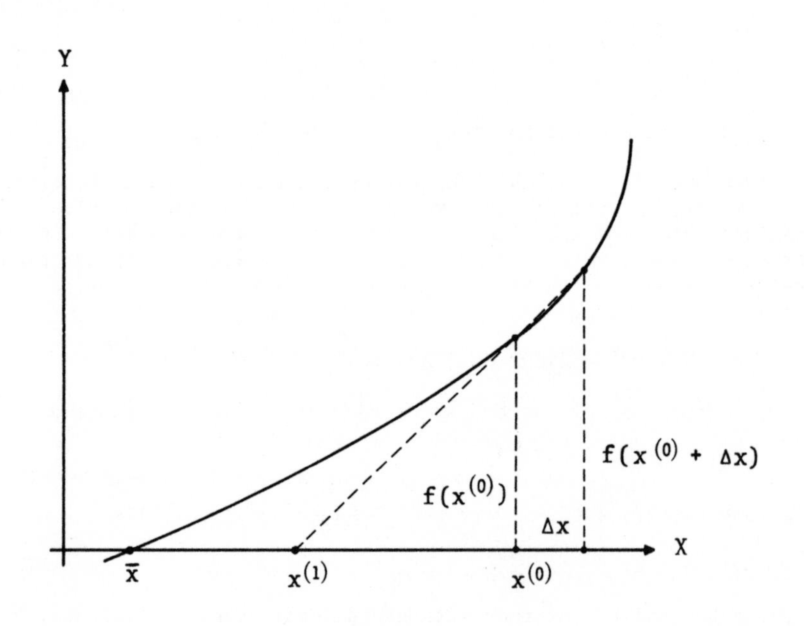

FIGURE 7.4

$$x^{(1)} = x^{(0)} - \frac{f(x^{(0)})}{\frac{f(x^{(0)}+\Delta x)-f(x^{(0)})}{\Delta x}} \tag{7.25a}$$

ow, just as one tried to improve on $x^{(0)}$ by generating $x^{(1)}$, one can try to mprove on $x^{(1)}$ by generating $x^{(2)}$. Thus, one finds the two points $(x^{(1)}, (x^{(1)}))$ and $(x^{(1)}+\Delta x, f(x^{(1)}+\Delta x))$ on the curve $y = f(x)$, draws the line hrough these two points, and, where it crosses the X-axis, calls the resulting ntercept $x^{(2)}$. As above, the formula for $x^{(2)}$ is

$$x^{(2)} = x^{(1)} - \frac{f(x^{(1)})}{\frac{f(x^{(1)}+\Delta x)-f(x^{(1)})}{\Delta x}} \tag{7.25b}$$

ext, one can generate $x^{(3)}$ from $x^{(2)}$ using

$$x^{(3)} = x^{(2)} - \frac{f(x^{(2)})}{\frac{f(x^{(2)}+\Delta x)-f(x^{(2)})}{\Delta x}}, \tag{7.25c}$$

nd then $x^{(4)}$ from $x^{(3)}$ using

$$x^{(4)} = x^{(3)} - \frac{f(x^{(3)})}{\frac{f(x^{(3)}+\Delta x)-f(x^{(3)})}{\Delta x}}, \tag{7.25d}$$

nd so forth.

he sequence of calculations described above is now given concisely as follows:

$$x^{(n+1)} = x^{(n)} - \frac{f(x^{(n)})}{\frac{f(x^{(n)}+\Delta x)-f(x^{(n)})}{\Delta x}}, \quad n = 0,1,2,\ldots . \tag{7.26}$$

n practice, the iteration formula (7.26), which is called Newton's formula, will ail if the denominator is zero. Geometrically, this happens when the straight ine through the two points $(x^{(n)}, f(x^{(n)}))$, $(x^{(n)}+\Delta x, f(x^{(n)}+\Delta x))$ is parallel to, nd hence cannot intersect, the X-axis. Also, in practice it is always advisable o carry out the division of the denominator terms in (7.26) first, since this enominator usually will be the quotient of two very small numbers. This particuar point will be illustrated in the example which follows. Finally, note that it s always advisable to choose Δx relatively small, which, in this book, will be aken be mean $\Delta x \sim 10^{-4}$.

n the event that the iteration (7.26) should fail to yield the desired approximaion, then one simply restarts it with a different $x^{(0)}$ and perhaps even a diferent Δx.

xample. Consider the problem of finding a root of (7.23), that is,

$$x - 10^{-x} = 0.$$

Setting

$$f(x) = x - 10^{-x}$$

implies

$$f(x+\Delta x) = (x+\Delta x) - 10^{-(x+\Delta x)}.$$

Thus,

$$\frac{f(x+\Delta x)-f(x)}{\Delta x} = 1 + 10^{-x}\left(\frac{1-10^{-\Delta x}}{\Delta x} \right).$$

Let us now fix Δx, say, at $\Delta x = 0.0001$, so that

$$\frac{1-10^{-\Delta x}}{\Delta x} = \frac{1-10^{-0.0001}}{0.0001} \sim 2.3023.$$

(The latter approximation can be obtained directly from any computer.) Newton's formula (7.26) then takes the specific form

$$x^{(n+1)} = x^{(n)} - \frac{x^{(n)}-10^{-x^{(n)}}}{1+(2.3023)10^{-x^{(n)}}}. \qquad (7.27)$$

From (7.27), beginning with $x^{(0)} = 0$, one has, to three decimal places,

$$x^{(1)} = 0.303$$

$$x^{(2)} = 0.394$$

$$x^{(3)} = 0.399$$

$$x^{(4)} = 0.399.$$

The iteration is stopped at $x^{(4)}$ because, to three decimal places, the iterates are no longer changing. Thus, the approximation to $\bar{x}$ is $x = 0.399$. Direct substitution of this value into (7.23) does not yield exactly zero, but yields

$$(0.399 - 10^{-0.399}) \sim 0.00001.$$

Greater accuracy can be obtained, if so desired, simply by carrying more decimal places during the iteration.

Incidentally, as done above, one should always verify, by direct substitution into the equation being solved, that the value of x to which one has converged is, indeed, an approximate solution. This is actually <u>necessary</u> because, at the start we had no idea whether or not, say, (7.23) even had a solution. Also, from the practical point of view, such a check provides one with a test of the computer program used, and even sometimes, of the reliability of the computer being used.

uppose next that one has two equations in two unknowns, say,

$$f_1(x,y) = 0$$

$$f_2(x,y) = 0.$$

hen we shall generalize Newton's formula (7.26) directly by the following coupled air of iteration formulas:

$$x^{(n+1)} = x^{(n)} - \frac{f_1(x^{(n)},y^{(n)})}{\dfrac{f_1(x^{(n)}+\Delta x,y^{(n)})-f_1(x^{(n)},y^{(n)})}{\Delta x}} \tag{7.28}$$

$$y^{(n+1)} = y^{(n)} - \frac{f_2(x^{(n+1)},y^{(n)})}{\dfrac{f_2(x^{(n+1)},y^{(n)}+\Delta y)-f_2(x^{(n+1)},y^{(n)})}{\Delta y}} . \tag{7.29}$$

ote that (7.29) makes use of $x^{(n+1)}$ as soon as it is generated by (7.28).

s an example, consider the system

$$y - 10^{-x} = 0$$

$$x - 10^{-y} = 0.$$

etting

$$f_1(x,y) = y - 10^{-x}$$

$$f_2(x,y) = x - 10^{-y}$$

nd $\Delta x = \Delta y = 0.0001$ implies, from (7.28) and (7.29), that

$$x^{(n+1)} = x^{(n)} - \frac{y^{(n)}-10^{-x^{(n)}}}{(2.3023)10^{-x^{(n)}}} \tag{7.30}$$

$$y^{(n+1)} = y^{(n)} - \frac{x^{(n+1)}-10^{-y^{(n)}}}{(2.3023)10^{-y^{(n)}}} . \tag{7.31}$$

ne now guesses both $x^{(0)}$ and $y^{(0)}$ to begin the iteration. Thus, for $x^{(0)}$ $y^{(0)} = 0$, (7.30) and (7.31) yield, to four decimal places,

$$x^{(1)} = 0 - \frac{0-1}{(2.3023)\cdot 1} = 0.4342$$

$$y^{(1)} = 0 - \frac{0.4342-1}{(2.3023)\cdot 1} = 0.2458,$$

and so the iteration would continue.

The iteration formulas (7.30) and (7.31) can be extended to much larger systems o equations in the fashion indicated, however, we will leave the specifics until su systems arise in a natural way, as they will in the next section.

7.7 AN ORBIT EXAMPLE

We turn now to an example of an orbit problem to illustrate the ideas developed thus far about gravitation. In the notation of Section 7.5, let $Gm_1 = 1$, thus "normalizing" the units to be used. For initial conditions, let $x_0 = 0.50$, $y_0 = 0.00$, $v_{0,x} = 0.00$, $v_{0,y} = 1.63$. Had we not changed the units of measuremen then numbers would have to be given with more astronomical magnitudes. From (7.3 (7.6) and (7.13) - (7.15), the equations of motion of the planet P can be writt in the form

$$x_{k+1} - x_k - \frac{\Delta t}{2}(v_{k+1,x}+v_{k,x}) = 0 \qquad (7.32$$

$$y_{k+1} - y_k - \frac{\Delta t}{2}(v_{k+1,y}+v_{k,y}) = 0 \qquad (7.33$$

$$v_{k+1,x} - v_{k,x} + \frac{(x_{k+1}+x_k)\Delta t}{(x_k^2+y_k^2)^{\frac{1}{2}}(x_{k+1}^2+y_{k+1}^2)^{\frac{1}{2}}[(x_k^2+y_k^2)^{\frac{1}{2}}+(x_{k+1}^2+y_{k+1}^2)^{\frac{1}{2}}]} = 0 \qquad (7.34$$

$$v_{k+1,y} - v_{k,y} + \frac{(y_{k+1}+y_k)\Delta t}{(x_k^2+y_k^2)^{\frac{1}{2}}(x_{k+1}^2+y_{k+1}^2)^{\frac{1}{2}}[(x_k^2+y_k^2)^{\frac{1}{2}}+(x_{k+1}^2+y_{k+1}^2)^{\frac{1}{2}}]} = 0 \qquad (7.35$$

For x_k, y_k, $v_{k,x}$ and $v_{k,y}$ known, equations (7.32) - (7.35) are four equations i the four unknowns x_{k+1}, y_{k+1}, $v_{k+1,x}$, $v_{k+1,y}$. If these are denoted, respectivel by

$$f_1(x_{k+1},y_{k+1},v_{k+1,x},v_{k+1,y}) = 0 \qquad (7.32$$

$$f_2(x_{k+1},y_{k+1},v_{k+1,x},v_{k+1,y}) = 0 \qquad (7.33$$

$$f_3(x_{k+1},y_{k+1},v_{k+1,x},v_{k+1,y}) = 0 \qquad (7.34$$

$$f_4(x_{k+1},y_{k+1},v_{k+1,x},v_{k+1,y}) = 0, \qquad (7.35$$

where

$$f_1(x_{k+1},y_{k+1},v_{k+1,x},v_{k+1,y}) = x_{k+1} - x_k - \frac{\Delta t}{2}(v_{k+1,x}+v_{k,x})$$

$$f_2(x_{k+1},y_{k+1},v_{k+1,x},v_{k+1,y}) = y_{k+1} - y_k - \frac{\Delta t}{2}(v_{k+1,y}+v_{k,y})$$

$$f_3 = v_{k+1,x} - v_{k,x} + \frac{(x_{k+1}+x_k)\Delta t}{(x_k^2+y_k^2)^{\frac{1}{2}}(x_{k+1}^2+y_{k+1}^2)^{\frac{1}{2}}[(x_k^2+y_k^2)^{\frac{1}{2}}+(x_{k+1}^2+y_{k+1}^2)^{\frac{1}{2}}]}$$

$$f_4 = v_{k+1,y} - v_{k,y} + \frac{(y_{k+1}+y_k)\Delta t}{(x_k^2+y_k^2)^{\frac{1}{2}}(x_{k+1}^2+y_{k+1}^2)^{\frac{1}{2}}[(x_k^2+y_k^2)^{\frac{1}{2}}+(x_{k+1}^2+y_{k+1}^2)^{\frac{1}{2}}]} .$$

hen, in analogy with (7.28) and (7.29), Newtonian iteration formulas to be sed for the solution of (7.32) - (7.35) are taken to be

$$x_{k+1}^{(n+1)} = x_{k+1}^{(n)} - \frac{f_1(x_{k+1}^{(n)},y_{k+1}^{(n)},v_{k+1,x}^{(n)},v_{k+1,y}^{(n)})}{\dfrac{f_1(x_{k+1}^{(n)}+\Delta x,y_{k+1}^{(n)},v_{k+1,x}^{(n)},v_{k+1,y}^{(n)})-f_1(x_{k+1}^{(n)},y_{k+1}^{(n)},v_{k+1,x}^{(n)},v_{k+1,y}^{(n)})}{\Delta x}} \tag{7.36}$$

$$y_{k+1}^{(n+1)} = y_{k+1}^{(n)} - \frac{f_2(x_{k+1}^{(n+1)},y_{k+1}^{(n)},v_{k+1,x}^{(n)},v_{k+1,y}^{(n)})}{\dfrac{f_2(x_{k+1}^{(n+1)},y_{k+1}^{(n)}+\Delta y,v_{k+1,x}^{(n)},v_{k+1,y}^{(n)})-f_2(x_{k+1}^{(n+1)},y_{k+1}^{(n)},v_{k+1,x}^{(n)},v_{k+1,y}^{(n)})}{\Delta y}} \tag{7.37}$$

$$v_{k+1,x}^{(n+1)} = v_{k+1,x}^{(n)} - \frac{f_3(x_{k+1}^{(n+1)},y_{k+1}^{(n+1)},v_{k+1,x}^{(n)},v_{k+1,y}^{(n)})}{\dfrac{f_3(x_{k+1}^{(n+1)},y_{k+1}^{(n+1)},v_{k+1,x}^{(n)}+\Delta v,v_{k+1,y}^{(n)})-f_3(x_{k+1}^{(n+1)},y_{k+1}^{(n+1)},v_{k+1,x}^{(n)},v_{k+1,y}^{(n)})}{\Delta v}} \tag{7.38}$$

$$v_{k+1,y}^{(n+1)} = v_{k+1,y}^{(n)} - \frac{f_4(x_{k+1}^{(n+1)},y_{k+1}^{(n+1)},v_{k+1,x}^{(n+1)},v_{k+1,y}^{(n)})}{\dfrac{f_4(x_{k+1}^{(n+1)},y_{k+1}^{(n+1)},v_{k+1,x}^{(n+1)},v_{k+1,y}^{(n)}+\Delta v)-f_4(x_{k+1}^{(n+1)},y_{k+1}^{(n+1)},v_{k+1,x}^{(n+1)},v_{k+1,y}^{(n)})}{\Delta v}} \tag{7.39}$$

ow, the equations (7.32) - (7.35) were ordered in the particular way given because hen the Newtonian iteration formulas (7.36) - (7.39) simplify immensely, as ollows. From (7.32) and (7.32'),

$$f_1(x_{k+1},y_{k+1},v_{k+1,x},v_{k+1,y}) = x_{k+1} - x_k - \frac{\Delta t}{2}(v_{k+1,x}+v_{k,x}),$$

so that

$$\frac{f_1(x_{k+1}+\Delta x, y_{k+1}, v_{k+1,x}, v_{k+1,y}) - f(x_{k+1}, y_{k+1}, v_{k+1,x}, v_{k+1,y})}{\Delta x} \equiv \frac{\Delta x}{\Delta x} \equiv 1.$$

In a similar fashion, all the denominators of the right-most terms in (7.36)-(7.39) are also unity. Hence (7.36)-(7.39) reduce to

$$x_{k+1}^{(n+1)} = x_{k+1}^{(n)} - f_1(x_{k+1}^{(n)}, y_{k+1}^{(n)}, v_{k+1,x}^{(n)}, v_{k+1,y}^{(n)}) \tag{7.36'}$$

$$y_{k+1}^{(n+1)} = y_{k+1}^{(n)} - f_2(x_{k+1}^{(n+1)}, y_{k+1}^{(n)}, v_{k+1,x}^{(n)}, v_{k+1,y}^{(n)}) \tag{7.37'}$$

$$v_{k+1,x}^{(n+1)} = v_{k+1,x}^{(n)} - f_3(x_{k+1}^{(n+1)}, y_{k+1}^{(n+1)}, v_{k+1,x}^{(n)}, v_{k+1,y}^{(n)}) \tag{7.38'}$$

$$v_{k+1,y}^{(n+1)} = v_{k+1,y}^{(n)} - f_4(x_{k+1}^{(n+1)}, y_{k+1}^{(n+1)}, v_{k+1,x}^{(n+1)}, v_{k+1,y}^{(n)}). \tag{7.39'}$$

In iterating with (7.36')-(7.39'), we begin, of course, with the given initial data x_0, y_0, $v_{0,x}$ and $v_{0,y}$. Then, each Newton iteration at time t_{k+1} is begun with the initial guess $x_{k+1}^{(0)} = x_k$, $y_{k+1}^{(0)} = y_k$, $v_{k+1,x}^{(0)} = v_{k,x}$, $v_{k+1,y}^{(0)} = v_{k,y}$.

As a typical example of the calculations, planetary motion was generated with $\Delta t = 0.001$ up to $t_{350000} = 350$. The total computing time was under 5 minutes on the UNIVAC 1108. There were 86+ orbits, the 86th of which is shown in Fig. 7.5. For this particular orbit, the period is $\tau = 4.05$ and the average of the absolute values of the x intercepts yields a semi-major axis length of 0.746.

The FORTRAN program for this example is given in Appendix C.

7.8 GRAVITY REVISITED

Let us show now that our theory of gravitation is so general that, from it, we can deduce the results we already know about gravity. From the discussion of Section 2.6, it follows that all we need do is show that the acceleration of gravity is given by (2.17), that is, $a_k \equiv -32$ ft/sec^2, $k = 0,1,2,\dots$. To do this, let the earth have mass m_1, and let it be positioned at the origin of an XY coordinate system, as shown in Fig. 7.6. Let particle P be positioned with its center on the X-axis, as shown in Fig. 7.6, and let its mass m_2 be relatively small compared to that of m_1, so that the motion of the earth due to gravitational forces can be neglected. Then, since the motion of P is along the X-axis, we have, by (7.13),

$$F_{k,x} = -\frac{Gm_1m_2(x_{k+1}+x_k)}{r_kr_{k+1}(r_k+r_{k+1})} = -\frac{Gm_1m_2(x_{k+1}+x_k)}{x_kx_{k+1}(x_k+x_{k+1})},$$

which implies

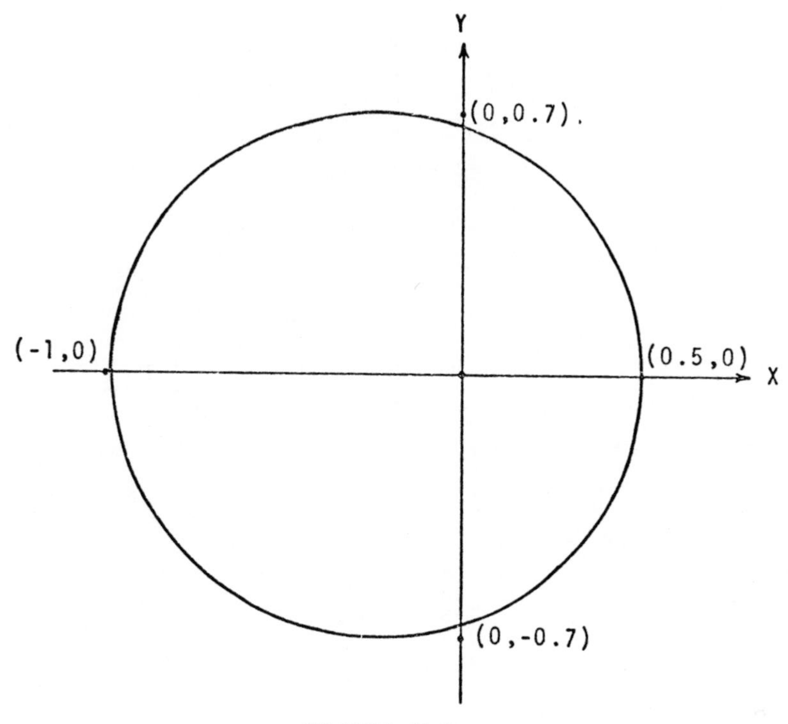

FIGURE 7.5

$$F_{k,x} = -\frac{Gm_1m_2}{x_kx_{k+1}} . \tag{7.40}$$

owever, one also has

$$F_{k,x} = m_2a_{k,x}, \tag{7.41}$$

o that, (7.40) and (7.41) imply

$$a_{k,x} = -\frac{Gm_1}{x_kx_{k+1}} . \tag{7.42}$$

ote, first, then, from (7.42), that the acceleration of P does not depend on the ass of P, but only on the mass of the earth.

ext, let us merely accept the astronomers' calculation of the mass of the earth, hich, in grams, is approximately

$$m_1 = (0.596)10^{28} \text{ gr} ,$$

nd also accept the usual estimate that the mean radius of the earth is 3959 miles. ecalling also (7.2), that is, that in cgs units $G = (6.67)10^{-8}$, we now have all

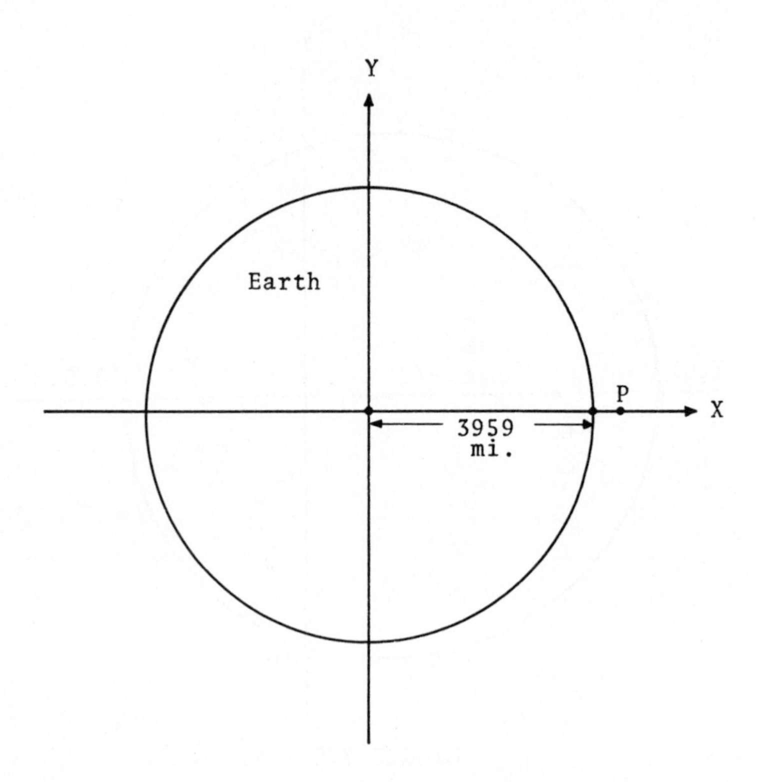

FIGURE 7.6

the data necessary to determine $a_{k,x}$ in (7.42). The only problem is that the various units are not consistent. So, if we merely note that to four decimal plac

$$.3936 \text{ in} = 1 \text{ cm},$$

then we can calculate $a_{k,x}$ in cgs units as follows.

From Fig. 7.6, one realizes quickly that the size of the earth is so great that i initially, P is close to the earth, then the distance P falls is very small compared to the radius of the earth. Thus, to a high degree of accuracy, $x_k = 3959$ mi, $k = 0,1,2,\ldots$, or, equivalently,

$$x_k = (6.373)10^8 \text{cm}, \quad k = 0,1,2,\ldots .$$

Substitution into (7.42) then yields

$$a_{k,x} = -\frac{(6.67)10^{-8}(0.598)10^{28}}{(6.373)^2 10^{16}} \sim -982 \text{ cm/sec}^2, \tag{7.43}$$

or, changing cm into feet,

$$a_{k,x} \sim -32 \text{ ft/sec}^2, \quad k = 0,1,2,\ldots \tag{7.44}$$

ich is the desired deduction.

7.9 ATTRACTION AND REPULSION

is worth showing now how one can extend the force law given in (7.12) - (7.15) include repulsion. This can be done by rewriting (7.13) and (7.14) as

$$F_{k,x} = -\frac{Gm_1m_2(x_{k+1}+x_k)}{r_kr_{k+1}(r_k+r_{k+1})} + \frac{Hm_1m_2\left[\sum_{j=0}^{m-2}(r_k^jr_{k+1}^{m-j-2})\right](x_{k+1}+x_k)}{r_k^{m-1}r_{k+1}^{m-1}(r_{k+1}+r_k)} \tag{7.45}$$

$$F_{k,y} = -\frac{Gm_1m_2(y_{k+1}+y_k)}{r_kr_{k+1}(r_k+r_{k+1})} + \frac{Hm_1m_2\left[\sum_{j=0}^{m-2}(r_k^jr_{k+1}^{m-j-2})\right](y_{k+1}+y_k)}{r_k^{m-1}r_{k+1}^{m-1}(r_{k+1}+r_k)}, \tag{7.46}$$

ere H is a constant of repulsion and $m \geq 2$. Intuitively, (7.45) - (7.46) ress attraction as a $1/r^2$ component and repulsion as a $1/r^m$ component, 2.

show that energy is conserved by (7.45) - (7.46), we need only establish a new ential function V_k and repeat the argument that led to (7.20). Since we know t the attracting portion of (7.45) and (7.46) is energy conserving, let us sim- show that the same is true of the repelling portion, which will establish con- vation. Hence, (7.7), (7.45) and (7.46) imply

$$\begin{aligned} W_n &= \sum_{k=0}^{n-1}\left\{(x_{k+1}-x_k)\frac{Hm_1m_2\left[\sum_{j=0}^{m-2}(r_k^jr_{k+1}^{m-j-2})\right](x_{k+1}+x_k)}{r_k^{m-1}r_{k+1}^{m-1}(r_{k+1}+r_k)}\right. \\ &\qquad \left. + (y_{k+1}-y_k)\frac{Hm_1m_2\left[\sum_{j=0}^{m-2}(r_k^jr_{k+1}^{m-j-2})\right](y_{k+1}+y_k)}{r_k^{m-1}r_{k+1}^{m-1}(r_{k+1}+r_k)}\right\} \\ &= Hm_1m_2\sum_{k=0}^{n-1}\left\{\frac{\left[\sum_{j=0}^{m-2}(r_k^jr_{k+1}^{m-j-2})\right]}{r_k^{m-1}r_{k+1}^{m-1}(r_{k+1}+r_k)}[x_{k+1}^2 - x_k^2 + y_{k+1}^2 - y_k^2]\right\} \\ &= Hm_1m_2\sum_{k=0}^{n-1}\left\{\frac{\left[\sum_{j=0}^{m-2}(r_k^jr_{k+1}^{m-j-2})\right](r_{k+1}-r_k)}{r_k^{m-1}r_{k+1}^{m-1}}\right\}. \end{aligned}$$

But,

$$\left[\sum_{j=0}^{m-2}(r_k^j r_{k+1}^{m-j-2})\right](r_{k+1}-r_k) \equiv r_{k+1}^{m-1} - r_k^{m-1},$$

so that

$$W_n = Hm_1m_2\sum_{k=0}^{n-1}\left(\frac{r_{k+1}^{m-1}-r_k^{m-1}}{r_k^{m-1}r_{k+1}^{m-1}}\right) = Hm_1m_2\sum_{k=0}^{n-1}\left(\frac{1}{r_k^{m-1}} - \frac{1}{r_{k+1}^{m-1}}\right)$$

$$= Hm_1m_2\left(\frac{1}{r_0^{m-1}} - \frac{1}{r_n^{m-1}}\right).$$

Defining the potential function V_k by

$$V_k = \frac{Hm_1m_2}{r_k^{m-1}}$$

yields

$$W_n = V_0 - V_n,$$

and conservation follows.

Finally, the latter discussion implies that (7.45) and (7.46) can be extended i[n] energy conserving fashion to include attraction as a $1/(r^n)$ component, $n \geq 2$ and repulsion as a $1/(r^m)$ component, $m \geq 2$, if the force is defined by

$$F_{k,x} = -\frac{Gm_1m_2\left[\sum_{j=0}^{n-2}(r_k^j r_{k+1}^{n-j-2})\right](x_{k+1}+x_k)}{r_k^{n-1}r_{k+1}^{n-1}(r_{k+1}+r_k)} + \frac{Hm_1m_2\left[\sum_{j=0}^{m-2}(r_k^j r_{k+1}^{m-j-2})\right](x_{k+1}+x_k)}{r_k^{m-1}r_{k+1}^{m-1}(r_{k+1}+r_k)} \quad (7.$$

$$F_{k,y} = -\frac{Gm_1m_2\left[\sum_{j=0}^{n-2}(r_k^j r_{k+1}^{n-j-2})\right](y_{k+1}+y_k)}{r_k^{n-1}r_{k+1}^{n-1}(r_{k+1}+r_k)} + \frac{Hm_1m_2\left[\sum_{j=0}^{m-2}(r_k^j r_{k+1}^{m-j-2})\right](y_{k+1}+y_k)}{r_k^{m-1}r_{k+1}^{m-1}(r_{k+1}+r_k)} \quad (7.$$

rces defined by (7.47)-(7.48) are fundamental in the classical approach to molecar dynamics (see, e.g., Hirschfelder, Curtis, and Bird), and will be applied in a gnificant fashion in later chapters.

7.10 REMARKS

his theoretical formulation of Mechanics, Newton made only three general physil assumptions. We have already alluded to two of these in Sections 3.3 and 3.4. btly underlying the nature of gravitational interaction, as described in Section 3, is the third assumption, which, for completeness, is now given as follows.

iom 3. Corresponding to the force of an action, there is a force of reaction ich is of the same magnitude but of opposite direction as the force of action.

th just a little thought, it can be seen also that Axiom 3 was, in fact, used in tting up the force formulas between particles of an elastic string in Chapter 5.

7.11 EXERCISES - CHAPTER 7

1 What is the difference between gravity and gravitation?

2 What is the meaning of the expression "The Copernican revolution"?

3 Design and execute an experiment to show that gravitational attraction exists.

4 For $\Delta t = 0.2$ and $t_k = k\Delta t$, describe the planar motion from t_0 to t_5 of a particle P of unit mass for each of the following cases:

(a) $\vec{r}_0 = (0,0)$, $\vec{v}_0 = (0,0)$, $\vec{F}_k = (1,0)$

(b) $\vec{r}_0 = (0,0)$, $\vec{v}_0 = (0,0)$, $\vec{F}_k = (5x_k, 5y_k)$

(c) $\vec{r}_0 = (-1,-1)$, $\vec{v}_0 = (10,10)$, $\vec{F}_k = (5x_k, 5y_k)$

(d) $\vec{r}_0 = (0,0)$, $\vec{v}_0 = (0,0)$, $\vec{F}_k = (1+y_k, 1-x_k)$

(e) $\vec{r}_0 = (-1,1)$, $\vec{v}_0 = (1,-1)$, $\vec{F}_k = (5y_k^2, -10x_k^2)$

(f) $\vec{r}_0 = (-1,-1)$, $\vec{v}_0 = (5,-5)$, $\vec{F}_k = (x_k^2 - y_k^2, -1)$

(g) $\vec{r}_0 = (1,2)$, $\vec{v}_0 = (1,-1)$, $\vec{F}_k = (x_k - y_k, x_k + y_k)$.

5 For $\Delta t = 0.2$ and $t_k = k\Delta t$, describe the three dimensional motion from t_0 to t_5 of a particle P of unit mass for each of the following cases:

(a) $\vec{r}_0 = (0,0,0)$, $\vec{v}_0 = (0,0,0)$, $\vec{F}_k = (0,10,0)$

(b) $\vec{r}_0 = (0,0,0)$, $\vec{v}_0 = (0,0,0)$, $\vec{F}_k = (5,0,-5)$

(c) $\vec{r}_0 = (1,1,1)$, $\vec{v}_0 = (0,0,0)$, $\vec{F}_k = (10z_k, 5y_k, -5x_k)$

(d) $\vec{r}_0 = (1,-1,1)$, $\vec{v}_0 = (1,0,-1)$, $\vec{F}_k = (z_k^2+x_k^2, z_k^2+y_k^2, x_k^2+y_k^2)$

(e) $\vec{r}_0 = (0,0,0)$, $\vec{v}_0 = (1,1,-1)$, $\vec{F}_k = (x_k^2+y_k+z_k, 1, -1)$

7.6 Find the kinetic energy at t_0 and at t_5 for each of the motions of Exercise 7.4.

7.7 Find the kinetic energy at t_0 and at t_5 for each of the motions of Exercise 7.5.

7.8 Verify Theorem 7.1 at t_5 for each of the motions of Exercise 7.4.

7.9 Verify Theorem 7.1 at t_5 for each of the motions of Exercise 7.5.

7.10 Try to find an approximate root for each of the following equations by Newton's method with $\Delta x = 0.0001$ and $x^{(0)} = 0$.

(a) $x - 7 = 0$

(b) $3x - 10^{-x} = 0$

(c) $x - 10^{-x} = 1$

(d) $x = \cos x$

(e) $x^2 - 2 = 0$

(f) $x^3 - 2 = 0$.

7.11 Find an approximate root for each equation of Exercise 7.10 by Newton's method with $\Delta x = 0.0001$ and with a choice of $x^{(0)}$ different from zero. Check your answers.

7.12 Try to find an approximate solution for each of the following systems by Newton's method with $\Delta x = \Delta y = 0.0001$, $x^{(0)} = y^{(0)} = 0$.

(a) $7x + y = 6$
$x - 7y = 8$

(b) $x^2 + y^2 = 25$
$x + 10y = -9$

(c) $x^2 + y^2 = 25$
$e^{-x} + e^y = 1$

(d) $10x + x \sin \pi y = 9$
$y \sin \pi x - 9y = 9$.

7.13 Find an approximate solution for each of the systems of equations in Exercise 7.12 by Newton's method with $\Delta x = \Delta y = 0.00001$. Check your answers.

.14 Describe the motion of a planet P whose dynamical equations are (7.32) - (7.35) for each of the following sets of initial conditions

(a) $x_0 = 0.50$, $y_0 = 0.00$, $v_{0,x} = 0.00$, $v_{0,y} = 1.80$

(b) $x_0 = 0.50$, $y_0 = 0.00$, $v_{0,x} = 0.00$, $v_{0,y} = 1.40$

(c) $x_0 = 0.50$, $y_0 = 0.00$, $v_{0,x} = 0.50$, $v_{0,y} = 1.80$

(d) $x_0 = 0.50$, $y_0 = 0.00$, $v_{0,x} = -0.50$, $v_{0,y} = 1.40$

(e) $x_0 = 0.50$, $y_0 = 0.50$, $v_{0,x} = -0.50$, $v_{0,y} = 0.50$.

.15 Show that the denominators of the right-most terms in each of (7.36) - (7.39) are unity.

.16 Show that, for $m \geq 2$,

$$\frac{Hm_1m_2\left[\sum_{j=0}^{m-2}(r_k^j r_{k+1}^{m-j-2})\right]}{r_k^{m-1}r_{k+1}^{m-1}}$$

behaves like $1/r^m$.

.17 Show that

$$\left[\sum_{j=0}^{m-2}(r_k^j r_{k+1}^{m-j-2})\right](r_{k+1}-r_k) \equiv r_{k+1}^{m-1} - r_k^{m-1}.$$

.18 Show that Axiom 3 of Section 7.10 is valid for adjacent pairs of string particles under force law (5.1) with tension formulas (5.8) and (5.9).

.19 Show that if the motion of a particle P is conservative then the force $\vec{F}_k$ and potential V_k at time t_k are always related by

$$\vec{F}_k = -\frac{(V_{k+1}-V_k)}{(r_{k+1}-r_k)} \cdot \frac{(\vec{r}_{k+1}+\vec{r}_k)}{(r_{k+1}+r_k)}, \quad r_k \neq r_{k+1}.$$

Verify this formula for the motions of Sections 2.6, 4.3, 7.5, and 7.9.

CHAPTER 8

The Three-Body Problem

8.1 INTRODUCTION

Our discussion of gravitation in Chapter 7 has left some very interesting and impo
tant matters unexplored. We did not study the possibilities that the two bodies had comparable masses, or that there might have been more than two bodies, or that there might have been other physical quantities conserved, not just energy. All these matters need to be studied if one wants a fuller understanding of, for example, the solar system.

The simplest, nontrivial problem in which all these considerations are important is the three-body problem, which is formulated as follows. Given the masses, posi tions and velocities of three particles, the problem is to determine their motions if each is under the gravitational influence of the other two. It is to this prob lem that attention is directed in the present chapter.

Note that the study of the motion of only two bodies of comparable masses can be explored by the methodology of Chapter 7 merely by taking a moving coordinate system whose origin is at the center of one of the bodies. In this coordinate system only one body is in motion. Another method of studying two body motion, as will become clear later, is to introduce a third body of <u>zero</u> mass and to solve a three body problem by the methodology of this chapter.

8.2 THE EQUATIONS OF MOTION

For $\Delta t > 0$ and $t_k = k\Delta t$, $k = 0,1,2,\ldots,$ and for each $i = 1,2,3,$ let particle P_i of mass m_i be located at $\vec{r}_{i,k} = (x_{i,k}, y_{i,k})$, have velocity $\vec{v}_{i,k} = (v_{i,k,x}, v_{i,k,y})$, and acceleration $\vec{a}_{i,k} = (a_{i,k,x}, a_{i,k,y})$ at time t_k. In analogy with (7.3) and (7.4), let

$$\frac{\vec{v}_{i,k+1}+\vec{v}_{i,k}}{2} = \frac{\vec{r}_{i,k+1}-\vec{r}_{i,k}}{\Delta t}, \quad i = 1,2,3; \quad k = 0,1,2,\ldots, \tag{8.1}$$

$$\vec{a}_{i,k} = \frac{\vec{v}_{i,k+1}-\vec{v}_{i,k}}{\Delta t}, \quad i = 1,2,3; \quad k = 0,1,2,\ldots \,. \tag{8.2}$$

f course, (8.1) - (8.2) differ from (7.3) - (7.4) only by the addition of the subscript i, which enables one to associate a given velocity and acceleration with a particular particle in the system.

o relate force and acceleration, we assume a discrete Newtonian equation

$$\vec{F}_{i,k} = m_i \vec{a}_{i,k}; \quad i = 1,2,3; \quad k = 0,1,2,\ldots, \tag{8.3}$$

here

$$\vec{F}_{i,k} = (F_{i,k,x}, F_{i,k,y}). \tag{8.4}$$

his time, the work W_n is defined for each positive integer n by

$$W_n = \sum_{i=1}^{3} W_{i,n}, \tag{8.5}$$

here

$$W_{i,n} = \sum_{k=0}^{n-1} [(x_{i,k+1} - x_{i,k})F_{i,k,x} + (y_{i,k+1} - y_{i,k})F_{i,k,y}]. \tag{8.6}$$

he exact derivation which yielded (7.11) implies, again,

$$W_i = \frac{m_i}{2}(v_{i,n,x}^2 + v_{i,n,y}^2) - \frac{m_i}{2}(v_{i,0,x}^2 + v_{i,0,y}^2), \tag{8.7}$$

o that if the kinetic energy $K_{i,k}$ of P_i at t_k is defined by

$$K_{i,k} = \frac{m_i}{2}(v_{i,k,x}^2 + v_{i,k,y}^2), \tag{8.8}$$

hen

$$W_{i,n} = K_{i,n} - K_{i,0}. \tag{8.9}$$

efining the kinetic energy K_k of the system at time t_k by

$$K_k = \sum_{i=1}^{3} K_{i,k} \tag{8.10}$$

ields, finally,

$$W_n = K_n - K_0. \tag{8.11}$$

ext, the precise structure of the force components of (8.4) is given as follows. f $r_{ij,k}$ is the distance between P_i and P_j at time t_k, then, in analogy ith (7.13) and (7.14), set

$$F_{1,k,x} = -\frac{Gm_1m_2[(x_{1,k+1}+x_{1,k})-(x_{2,k+1}+x_{2,k})]}{r_{12,k}r_{12,k+1}(r_{12,k}+r_{12,k+1})}$$

$$-\frac{Gm_1m_3[(x_{1,k+1}+x_{1,k})-(x_{3,k+1}+x_{3,k})]}{r_{13,k}r_{13,k+1}(r_{13,k}+r_{13,k+1})} \quad (8.12)$$

$$F_{2,k,x} = -\frac{Gm_1m_2[(x_{2,k+1}+x_{2,k})-(x_{1,k+1}+x_{1,k})]}{r_{12,k}r_{12,k+1}(r_{12,k}+r_{12,k+1})}$$

$$-\frac{Gm_2m_3[(x_{2,k+1}+x_{2,k})-(x_{3,k+1}+x_{3,k})]}{r_{23,k}r_{23,k+1}(r_{23,k}+r_{23,k+1})} \quad (8.13)$$

$$F_{3,k,x} = -\frac{Gm_1m_3[(x_{3,k+1}+x_{3,k})-(x_{1,k+1}+x_{1,k})]}{r_{13,k}r_{13,k+1}(r_{13,k}+r_{13,k+1})}$$

$$-\frac{Gm_2m_3[(x_{3,k+1}+x_{3,k})-(x_{2,k+1}+x_{2,k})]}{r_{23,k+1}r_{23,k}(r_{23,k}+r_{23,k+1})}, \quad (8.14)$$

while $F_{1,k,y}$, $F_{2,k,y}$, $F_{3,k,y}$ are defined by interchanging y and x in (8.12), (8.13) and (8.14).

Each of (8.12) - (8.14) has two gravitational terms on the right-hand side because each particle is now influenced by two other particles.

8.3 CONSERVATION OF ENERGY

To establish the conservation of energy, consider, again (8.5). Substitution of (8.12) - (8.14) and the corresponding formulas for $F_{1,k,y}$, $F_{2,k,y}$ and $F_{3,k,y}$ into (8.5) yields readily

$$W_n = -Gm_1m_2 \sum_{k=0}^{n-1} \left(\frac{r_{12,k+1}-r_{12,k}}{r_{12,k}r_{12,k+1}} \right) - Gm_1m_3 \sum_{k=0}^{n-1} \left(\frac{r_{13,k+1}-r_{13,k}}{r_{13,k}r_{13,k+1}} \right)$$

$$- Gm_2m_3 \sum_{k=0}^{n-1} \left(\frac{r_{23,k+1}-r_{23,k}}{r_{23,k}r_{23,k+1}} \right)$$

$$= -Gm_1m_2\left(\frac{1}{r_{12,0}} - \frac{1}{r_{12,n}}\right) - Gm_1m_3\left(\frac{1}{r_{13,0}} - \frac{1}{r_{13,n}}\right)$$

$$- Gm_2m_3\left(\frac{1}{r_{23,0}} - \frac{1}{r_{23,n}}\right).$$

Defining the potential energy $V_{ij,k}$ of the pair P_i and P_j at t_k by

$$V_{ij,k} = -G\frac{m_i m_j}{r_{ij,k}}$$

mplies then that

$$W = V_{12,0} + V_{13,0} + V_{23,0} - V_{12,n} - V_{13,n} - V_{23,n}. \qquad (8.15)$$

f the potential energy V_k of the system at time t_k is defined by

$$V_k = V_{12,k} + V_{13,k} + V_{23,k},$$

hen (8.15) implies

$$W_n = V_0 - V_n. \qquad (8.16)$$

inally, elimination of W_n between (8.11) and (8.16) yields the desired result:

$$K_n + V_n = K_0 + V_0.$$

8.4 SOLUTION OF THE DISCRETE THREE-BODY PROBLEM

hough we now can describe a general algorithm for generating a solution of the hree-body problem, and, indeed, a FORTRAN program for the general problem is given n Appendix D, it is more instructive at present to consider a particular problem nd to show, in detail, how to solve it. The reasoning required for other problems s entirely analogous. Consider, therefore, as shown in Fig. 8.1, three particles $_1$, P_2, P_3, of equal masses which are normalized so that

$$m_1 = m_2 = m_3 = 10, \quad G = 1. \qquad (8.17)$$

et the initial positions and velocities be given by $x_{1,0} = 0$, $y_{1,0} = 100$, $_{2,0} = 100$, $y_{2,0} = 0$, $x_{3,0} = -100$, $y_{3,0} = 0$, $v_{1,0,x} = 0$, $v_{1,0,y} = -10$,

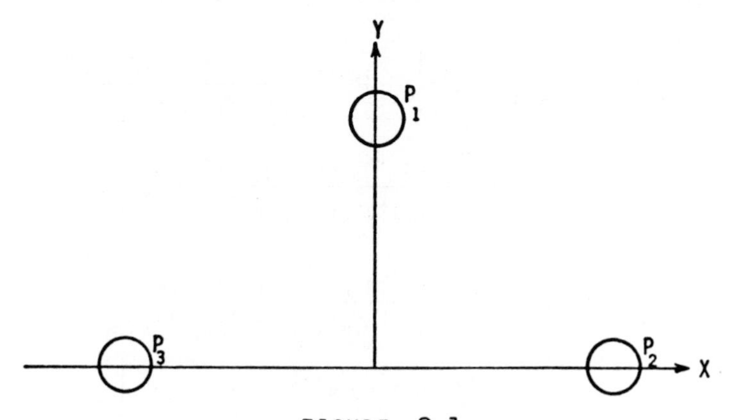

FIGURE 8.1

$v_{2,0,x} = -10$, $v_{2,0,y} = 0$, $v_{3,0,x} = 9.9$, $v_{3,0,y} = 0$. From (8.1) - (8.4), one ca as in (7.32) - (7.35), rewrite the equations of motion as follows:

$$x_{i,k+1} - x_{i,k} - \frac{\Delta t}{2}(v_{i,k+1,x}+v_{i,k,x}) = 0, \quad i = 1,2,3 \tag{8.18}$$

$$y_{i,k+1} - y_{i,k} - \frac{\Delta t}{2}(v_{i,k+1,y}+v_{i,k,y}) = 0, \quad i = 1,2,3 \tag{8.19}$$

$$v_{1,k+1,x} - v_{1,k,x} + 10\Delta t\left\{\frac{(x_{1,k+1}+x_{1,k})-(x_{2,k+1}+x_{2,k})}{r_{12,k}r_{12,k+1}(r_{12,k}+r_{12,k+1})} + \frac{(x_{1,k+1}+x_{1,k})-(x_{3,k+1}+x_{3,k})}{r_{13,k}r_{13,k+1}(r_{13,k}+r_{13,k+1})}\right\} = 0 \tag{8.20}$$

$$v_{1,k+1,y} - v_{1,k,y} + 10\Delta t\left\{\frac{(y_{1,k+1}+y_{1,k})-(y_{2,k+1}+y_{2,k})}{r_{12,k}r_{12,k+1}(r_{12,k}+r_{12,k+1})} + \frac{(y_{1,k+1}+y_{1,k})-(y_{3,k+1}+y_{3,k})}{r_{13,k}r_{13,k+1}(r_{13,k}+r_{13,k+1})}\right\} = 0 \tag{8.21}$$

$$v_{2,k+1,x} - v_{2,k,x} + 10\Delta t\left\{\frac{(x_{2,k+1}+x_{2,k})-(x_{1,k+1}+x_{1,k})}{r_{12,k}r_{12,k+1}(r_{12,k}+r_{12,k+1})} + \frac{(x_{2,k+1}+x_{2,k})-(x_{3,k+1}+x_{3,k})}{r_{23,k}r_{23,k+1}(r_{23,k}+r_{23,k+1})}\right\} = 0 \tag{8.22}$$

$$v_{2,k+1,y} - v_{2,k,y} + 10\Delta t\left\{\frac{(y_{2,k+1}+y_{2,k})-(y_{1,k+1}+y_{1,k})}{r_{12,k}r_{12,k+1}(r_{13,k}+r_{13,k+1})} + \frac{(y_{2,k+1}+y_{2,k})-(y_{3,k+1}+y_{3,k})}{r_{23,k}r_{23,k+1}(r_{23,k}+r_{23,k+1})}\right\} = 0 \tag{8.23}$$

$$v_{3,k+1,x} - v_{3,k,x} + 10\Delta t\left\{\frac{(x_{3,k+1}+x_{3,k})-(x_{1,k+1}+x_{1,k})}{r_{13,k}r_{13,k+1}(r_{13,k}+r_{13,k+1})} + \frac{(x_{3,k+1}+x_{3,k})-(x_{2,k+1}+x_{2,k})}{r_{23,k+1}r_{23,k}(r_{23,k}+r_{23,k+1})}\right\} = 0 \tag{8.24}$$

$$v_{3,k+1,y} - v_{3,k,y} + 10\Delta t\left\{\frac{(y_{3,k+1}+y_{3,k})-(y_{1,k+1}+y_{1,k})}{r_{13,k}r_{13,k+1}(r_{13,k}+r_{13,k+1})} + \frac{(y_{3,k+1}+y_{3,k})-(y_{2,k+1}+y_{2,k})}{r_{23,k+1}r_{23,k}(r_{23,k}+r_{23,k+1})}\right\} = 0 \tag{8.25}$$

here

$$r_{ij,k} = [(x_{i,k}-x_{j,k})^2 + (y_{i,k}-y_{j,k})^2]^{\frac{1}{2}}. \tag{8.26}$$

he solution of the twelve equation (8.18) - (8.25) for the twelve unknowns $x_{i,k+1}$, $_{i,k+1}$, $v_{i,k+1,x}$, $v_{i,k+1,y}$, $i = 1,2,3$, for each value of $k = 0,1,2,\ldots,$ rom the initial data is found by Newton's method with initial guess $x^{(0)}_{i,k+1} = x_{i,k}$, $^{(0)}_{i,k+1} = y_{i,k}$, $v^{(0)}_{i,k+1,x} = v_{i,k,x}$, $v^{(0)}_{i,k+1,y} = v_{i,k,y}$. In Fig. 8.2 are shown for $t = 0.1$ the deflections of the particles from times t_{95} to t_{105}. The motion f each particle is shown separately and the labels P_i, $i = 1,2,3$ are affixed at heir positions corresponding to t_{95}. The running time for one thousand time steps as under twenty seconds on the UNIVAC 1108.

8.5 THE OSCILLATORY NATURE OF PLANETARY PERIHELION MOTION

he methodology developed in Section 8.4 allows us to perform various types of comuter experiments related to planetary motion. In this section we will show how a ariation of the masses in a three-body problem can lead to the magnification of efined aspects of orbital behavior, thus making it easier to discover their preence. To do this, we will study several examples of what is known as perihelion otion. This type of motion is also of great interest in the Theory of General elativity.

n each example which follows the time step is $\Delta t = 0.001$ and CGS units are used, o that $G = (6.67)10^{-8}$.

xample 1. Consider the three-body problem for particles P_1, P_2 and P_3 with the ollowing initial data:

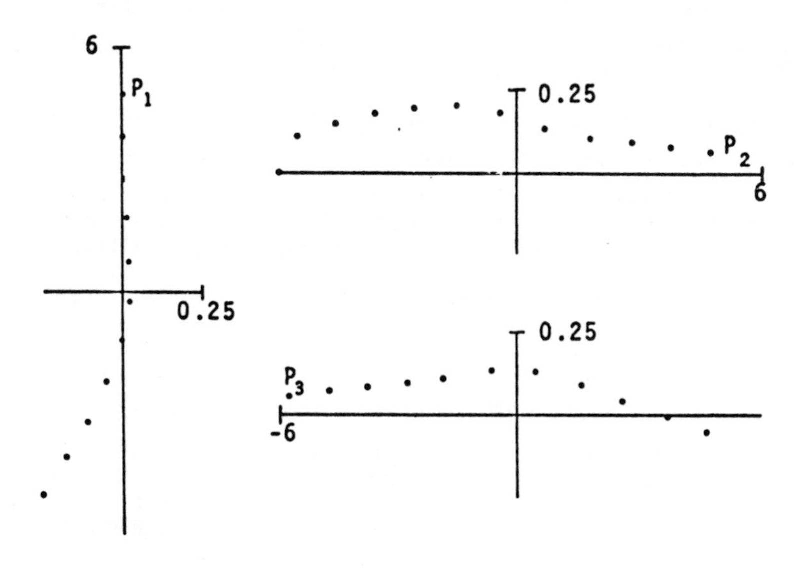

FIGURE 8.2

$m_1 = (6.67)^{-1}10^8$, $m_2 = (6.67)^{-1}10^6$, $m_3 = (6.67)^{-1}10^5$

$x_{1,0} = 0$, $x_{2,0} = 0.5$, $x_{3,0} = -1$

$y_{1,0} = 0$, $y_{2,0} = 0$, $y_{3,0} = 8$

$v_{1,0,x} = 0$, $v_{2,0,x} = 0$, $v_{3,0,x} = 0$

$v_{1,0,y} = 0$, $v_{2,0,y} = 1.63$, $v_{3,0,y} = -3.75$.

In the absence of P_3, the motion of P_2 relative to P_1 is the periodic orbit shown in Fig. 8.3, for which the period is $\tau = 3.901$. If the major axis of the motion is the line of greatest distance between any two points of an orbit, and i the length of the major axis is defined to be $2a$, then the major axis of P_2's motion relative to P_1 lies on the X-axis and $a = 0.730$. Incidentally, this orbit was constructed by solving the three body problem with $m_3 = 0$.

The initial data for P_3 were chosen so that this particle begins its motion at relatively large distance from both P_1 and P_2, arrives in the vicinity of $(-1,0)$ almost simultaneously with P_2, and proceeds past $(-1,0)$ at a relative high velocity, assuring only a short period of strong gravitational attraction. Particles P_2 and P_3 come closest in the third quadrant at t_{2125}, when P_2 at $(-0.9296,-0.1108)$ and P_3 is at $(-0.9325,-0.1012)$. The effect of the inte action is to deflect P_2 outward, as is seen clearly in Fig. 8.4, where the moti of P_2, relative to P_1, has been plotted from t_0 to t_{5000}, with the intege labels $n = 0,1,2,3,4,5$, marking the positions t_{1000n}. After having been defle ted, P_2 goes into the new orbit about P_1 which is shown in Fig. 8.5. The end points of the new major axis are $(0.4943, 0.1664)$ and $(-0.9105,-0.3075)$, so that $a = 0.74135$. The new period is $\tau = 3.9905$.

Now, the <u>perihelion</u> point of any orbit of P_2 about P_1 is its position which i closest to P_1 during that orbit. Since P_2 has been deflected into a new orbi its perihelion point has moved. The perihelion motion is measured by the angle o inclination θ of the new major axis with the X-axis, and is given by $\tan\theta = 0.34$. Note that the perihelion motion of this example is positive.

<u>Example 2</u>. The data of Example 1 were changed only for P_3 by setting $x_{3,0} = -0.5$, $y_{3,0} = 8.0$, $v_{3,0,x} = -0.25$, $v_{3,0,y} = -4.00$. This time the strongest gravitational effect between P_2 and P_3 occurs in the second quadrant at t_{196} when P_2 is at $(-0.94582, 0.01950)$ and p_3 is at $(-0.94418, 0.01796)$, and is perturbed into the new orbit shown in Fig. 8.6. The end points of the new maj axis are $(0.50724, -0.18349)$ and $(-0.92692, 0.33474)$, so that $a = 0.76246$, and the new period is $\tau = 4.162$. The resulting perihelion motion is now negativ since the angle θ of the new major axis with the X-axis is given by $\tan\theta = -0.36$.

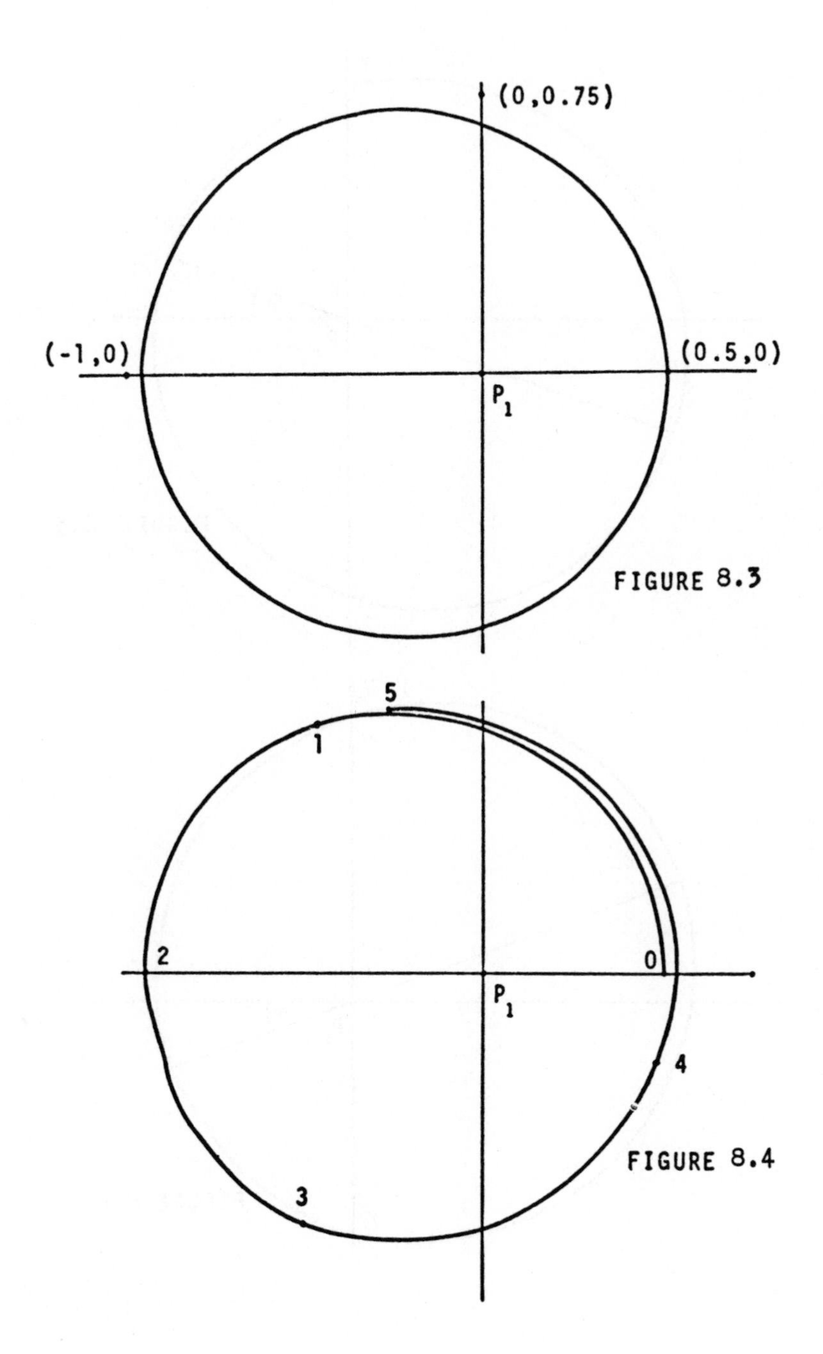

FIGURE 8.3

FIGURE 8.4

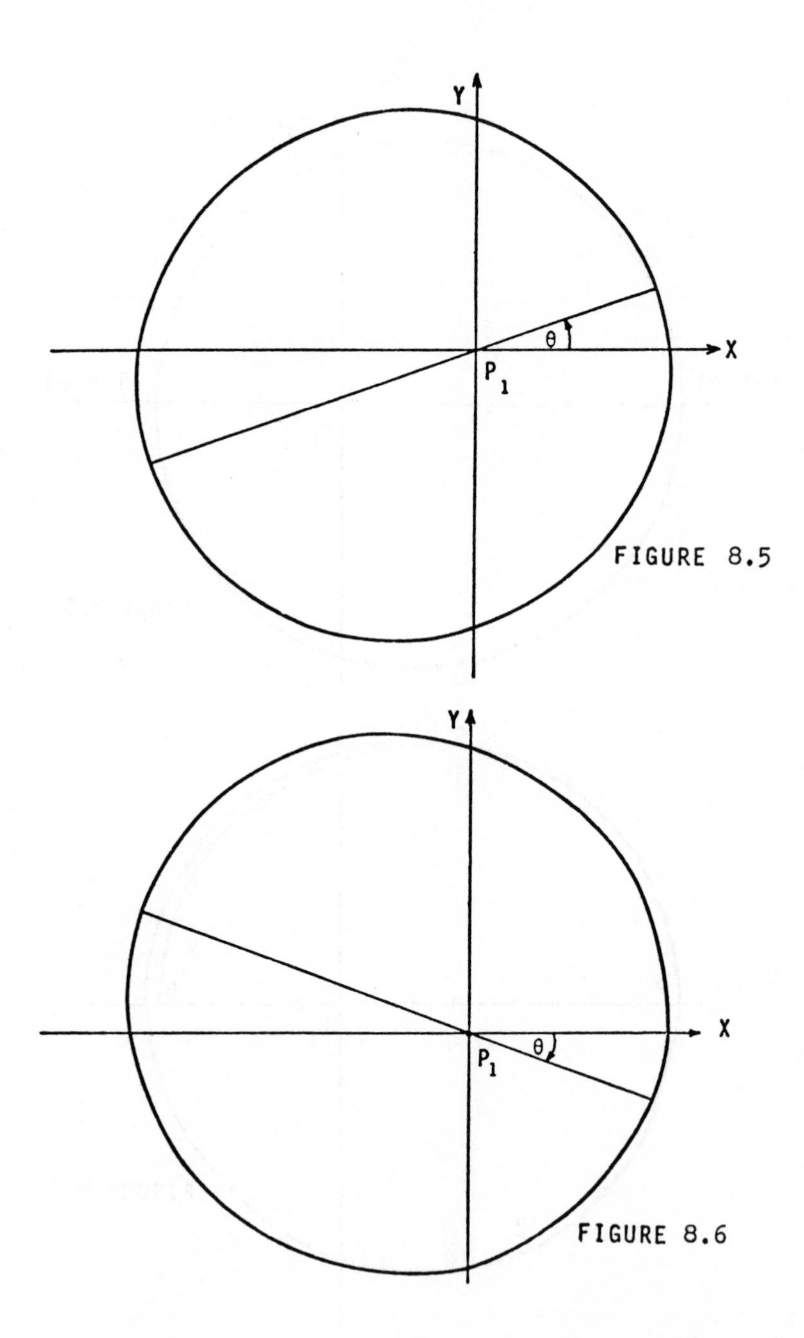

FIGURE 8.5

FIGURE 8.6

rom the above and similar examples, it follows that the major axis of P_2 is eflected in the same direction as is P_2. In actual planetary motions, as, for xample, in a Sun-Mercury-Venus system, where the mass of the sun is distinctly ominant, it can be concluded that when Mercury and Venus are relatively close in he first or in the third quadrants, the perihelion motion of Mercury must be perurbed a very small amount in the positive angular direction, while relative closeess in the second or in the fourth quadrants must result in a very small negative ngular perturbation. All such possibilities can occur for the motions of Mercury nd Venus. Thus, the perihelion motion of Mercury is a complex, nonlinear, oscilatory motion. These conclusions were verified on the computer with ten full orbits f Mercury. Most astronomy books leave the reader with the incorrect impression hat the perihelion motion of Mercury is uniform and always positive.

8.6 CENTER OF GRAVITY

n the remainder of this chapter, we will study various interesting and important roperties of three-body systems. These properties are not possessed by each individual particle, but are possessed by the system as a whole. We know, already, for xample, that a three-body system is conservative. But, no individual particle's otion need be conservative.

convenient place to begin is with the concept of the center of gravity. We know, or example, that if several weights are put on a light metal plate, then there is single point at which the entire set of weights can be balanced. This point is alled the center of gravity of the system because, if we support the system at his point, nothing will fall. In a three-body system, the center of gravity is efined formally as follows. Extension to more complex systems follows in the ndicated, natural way, simply by letting the number of particles be arbitrary, ather than three.

t time t_k, let P_i of mass m_i be at $(x_{i,k},y_{i,k})$, $i = 1,2,3$. Let

$$M = m_1 + m_2 + m_3$$

e the mass of the system. Then the unique point $(\overline{x}_k,\overline{y}_k)$ such that

$$\begin{aligned} M\overline{x}_k &= m_1x_{1,k} + m_2x_{2,k} + m_3x_{3,k} \\ M\overline{y}_k &= m_1y_{1,k} + m_2y_{2,k} + m_3y_{3,k} \end{aligned} \tag{8.27}$$

s called the center of gravity, or of mass, of the system at time t_k.

xample. At time t_7, let P_1 of mass 1 gram, P_2 of mass 2 grams and P_3 f mass 3 grams be at $(x_{1,7},y_{1,7})$, $(x_{2,7},y_{2,7})$, $(x_{3,7},y_{3,7})$, respectively. f $x_{1,7} = 12$, $y_{1,7} = 6$, $x_{2,7} = 3$, $y_{2,7} = -12$, $x_{3,7} = -4$, $y_{3,7} = 0$, then the enter of gravity $(\overline{x}_7,\overline{y}_7)$ at t_7 is given by

$$\overline{x}_7 = \frac{1}{6}[1\cdot12 + 2\cdot3 + 3\cdot(-4)] = 1$$

$$\overline{y}_7 = \frac{1}{6}[1\cdot6 + 2\cdot(-12) + 3\cdot0] = -3.$$

Let us see next if we can deduce what the motion of the center of gravity of a three-body problem must be. Let $t_k = k\Delta t$, $k = 0,1,2,\ldots$. Then, from (8.3) and (8.12) - (8.13) one has

$$m_1 a_{1,k,x} + m_2 a_{2,k,x} + m_3 a_{3,k,x} = 0, \quad k \geq 0. \tag{8.28}$$

Hence,

$$m_1(v_{1,k+1,x}-v_{1,k,x}) + m_2(v_{2,k+1,x}-v_{2,k,x}) + m_3(v_{3,k+1,x}-v_{3,k,x}) = 0. \tag{8.29}$$

Summing both sides of (8.29) over k from 0 to $j-1$, where $j \geq 1$, yields

$$m_1(v_{1,j,x}-v_{1,0,x}) + m_2(v_{2,j,x}-v_{2,0,x}) + m_3(v_{3,j,x}-v_{3,0,x}) = 0. \tag{8.30}$$

However, since (8.30) is valid even if $j = 0$, it follows that

$$m_1 v_{1,j,x} + m_2 v_{2,j,x} + m_3 v_{3,j,x} = c_1, \quad j \geq 0 \tag{8.31}$$

where

$$c_1 = m_1 v_{1,0,x} + m_2 v_{2,0,x} + m_3 v_{3,0,x}. \tag{8.32}$$

Since (8.31) is valid for any j, it must be valid if j is replaced by $j+1$, so that

$$m_1 v_{1,j+1,x} + m_2 v_{2,j+1,x} + m_3 v_{3,j+1,x} = c_1. \tag{8.33}$$

Addition of (8.31) and (8.33) then yields

$$m_1\left(\frac{v_{1,j+1,x}+v_{1,j,x}}{2}\right) + m_2\left(\frac{v_{2,j+1,x}+v_{2,j,x}}{2}\right) + m_3\left(\frac{v_{3,j+1,x}+v_{3,j,x}}{2}\right) = c_1$$

or, equivalently,

$$m_1(x_{1,j+1}-x_{1,j}) + m_2(x_{2,j+1}-x_{2,j}) + m_3(x_{3,j+1}-x_{3,j}) = c_1\Delta t, \quad j \geq 0. \tag{8.34}$$

Summing both sides of (8.34) with respect to j from 0 to $n-1$, for $n \geq 1$, yields

$$m_1(x_{1,n}-x_{1,0}) + m_2(x_{2,n}-x_{2,0}) + m_3(x_{3,n}-x_{3,0}) = c_1 t_n. \tag{8.35}$$

However, (8.35) is valid also for $n = 0$, so that

$$m_1 x_{1,n} + m_2 x_{2,n} + m_3 x_{3,n} = c_1 t_n + c_2, \quad n \geq 0, \tag{8.36}$$

where

$$c_2 = m_1 x_{1,0} + m_2 x_{2,0} + m_3 x_{3,0}. \tag{8.37}$$

In a fashion analogous to the derivation of (8.36), it follows also that

$$m_1y_{1,n} + m_2y_{2,n} + m_3y_{3,n} = d_1t_n + d_2, \quad n \geq 0, \tag{8.38}$$

where

$$d_1 = m_1v_{1,0,y} + m_2v_{2,0,y} + m_3v_{3,0,y}$$

$$d_2 = m_1y_{1,0} + m_2y_{2,0} + m_3y_{3,0}.$$

Hence, (8.36) and (8.37) imply

$$M\bar{x}_n = c_1t_n + c_2, \quad n \geq 0$$

$$M\bar{y}_n = d_1t_n + d_2, \quad n \geq 0,$$

from which it follows that the motion of the center of gravity moves along a straight line if $c_1^2 + d_1^2 \neq 0$ and is fixed if $c_1^2 + d_1^2 = 0$.

8.7 CONSERVATION OF LINEAR MOMENTUM

In studying the motion of an object, it is convenient to have a measure of how much force it takes to stop the object from its motion. The measure is called the object's linear momentum and is defined as follows. If particle P has mass m and velocity $\vec{v}$, then its linear momentum is defined as $m\vec{v}$.

If, instead of one particle, there are more, say, three particles, then the linear momentum of the system is defined as the sum of the linear momenta of the three particles.

If one now reexamines the discussion of the motion of the center of gravity in Section 8.6, one sees that one can deduce an interesting result from the equations developed there. Equation (8.31), and the corresponding equation for the y components, is a precise statement that at any time, the linear momentum of the system is always the same as what it was at t_0. Thus, the motion of the system, however complex it may be, always conserves the sum of the linear momenta of its particles. This result is our second basic conservation law and is called the Law of Conservation of Linear Momentum.

The third, and final, conservation law, the Conservation of Angular Momentum, will be developed next.

8.8 CONSERVATION OF ANGULAR MOMENTUM

Let us anticipate now some simple ideas about rotating systems which will be developed in greater detail in Chapter 11. For the present, let us make some simple observations about rotating bicycle wheels.

Suppose a bicycle wheel is set in a horizontal position and is made to rotate around its axle. Then a very interesting effect results, and it is most noticeable when the wheel is rotating very quickly. If one exerts a force perpendicular to the plane of the wheel, then the rotating wheel tends to resist this force. It seems to want to continue to rotate in its plane, and even though its motion is entirely two dimensional, it seems to push back on any force which is not in this plane of motion. Moreover, the greater its mass, or its radius, or its speed, the more marked is this tendency to resist. We say that this effect is due to its angular momentum and we will need to develop a formula for measuring this momentum.

The same effect due to angular momentum is also noticeable when one ties a metal ball to the end of a string of length r and, by twirling, makes it rotate quickly in a plane, circular path of radius r. The ball resists any force perpendicular to its plane of motion. Indeed, the same effect is also apparent in planetary motion, since, for example, one can think of the above ball replaced by the earth and the string constraint replaced by the gravitational attraction with the sun.

Let us see then how a formula for angular momentum can be developed. Consider, for simplicity, first a particle P of mass m in rotation at a constant speed on a circle 0 of radius r, as shown in Fig. 8.7. From the above discussion, we know that, whatever else, the magnitude of the angular momentum of P should vary directly with its mass, its distance r to the center, and with its speed. It has been found that a very convenient way to do this is to define angular momentum as the vector $\vec{L}$ which is given as follows:

$$\vec{L} = m(\vec{r} \times \vec{v}). \tag{8.39}$$

The reason for this is that, by (6.33),

$$|\vec{L}| = m|\vec{r}|\,|\vec{v}|\sin\theta.$$

But, for circular motion, as shown in Fig. 8.7, $\theta = \pi/2$, so that

$$|\vec{L}| = m|\vec{r}|\,|\vec{v}|,$$

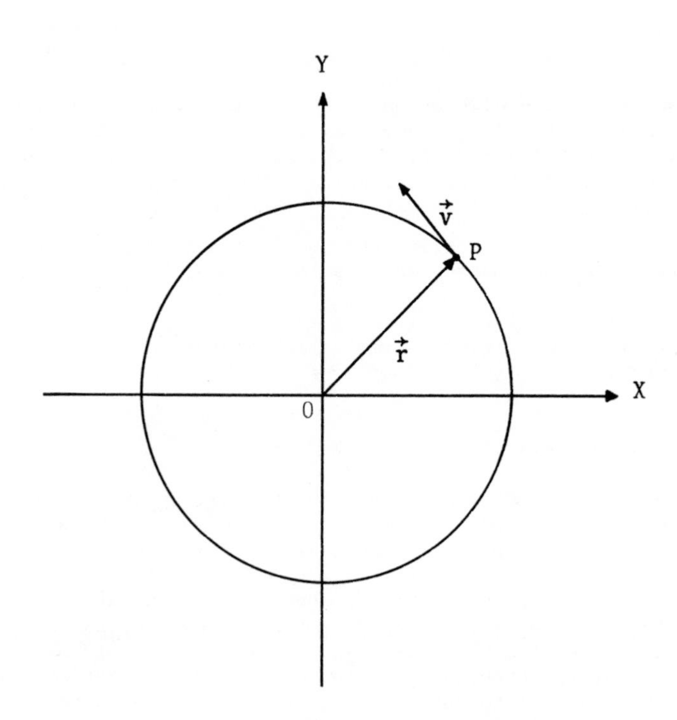

FIGURE 8.7

hich is precisely the kind of formula we want since $|\vec{L}|$ varies directly with ass, distance, and speed.

ut, now that the discussion is getting precise, one can argue that the analogy bove, of the earth in rotation about the sun, is not really a good one since the arth's motion is not circular. However, it is very nearly circular, and in order o incorporate these kinds of motions also into our definition, we simply proceed s follows. At t_k, let particle P of mass m be at $\vec{r}_k$ and have velocity $\vec{v}_k$. hen the particle's angular momentum $\vec{L}(t_k) = \vec{L}_k$ is defined by

$$\vec{L}_k = m(\vec{r}_k \times \vec{v}_k). \tag{8.40}$$

f $|\vec{r}_k|$ and $|\vec{v}_k|$ do not vary with time, then (8.40) is the same as (8.39). If he motion is not circular but, say, is close to circular, then, from (6.33)

$$|\vec{L}_k| = m|\vec{r}_k||\vec{v}_k|\sin\theta, \tag{8.41}$$

here θ is close to $\pi/2$.

et us show next that if one determines the angular momentum of each particle of a hree-body problem, then, amazingly enough, the sum of these three angular momenta ever changes. To do this, at time t_j let particle P_i of mass m_i be located t $\vec{r}_{i,j}$, have velocity $\vec{v}_{i,j}$, and have angular momentum $\vec{L}_{i,j}$, that is,

$$\vec{L}_{i,j} = m_i(\vec{r}_{i,j} \times \vec{v}_{i,j}). \tag{8.42}$$

n the system of three particles, let the system angular momentum $\vec{L}_j$ at t_j be efined by

$$\vec{L}_j = \sum_{i=1}^{3} \vec{L}_{i,j}. \tag{8.43}$$

hat we wish to show is that

$$\vec{L}_j = \vec{L}_0, \quad j = 1,2,3,\ldots, \tag{8.44}$$

nd this is done as follows.

rom (8.42) and the laws of vector cross products,

$$\begin{aligned}\vec{L}_{i,k+1} - \vec{L}_{i,k} &= m_i(\vec{r}_{i,k+1}\times\vec{v}_{i,k+1}) - m_i(\vec{r}_{i,k}\times\vec{v}_{i,k}) \\ &= m_i\left[(\vec{r}_{i,k+1}-\vec{r}_{i,k})\times\left(\frac{\vec{v}_{i,k+1}+\vec{v}_{i,k}}{2}\right)\right. \\ &\qquad \left. + \left(\frac{\vec{r}_{i,k+1}+\vec{r}_{i,k}}{2}\right)\times(\vec{v}_{i,k+1}-\vec{v}_{i,k})\right].\end{aligned}$$

rom (8.1) - (8.4), then,

$$\vec{L}_{i,k+1} - \vec{L}_{i,k} = m_i\left[(\vec{r}_{i,k+1}-\vec{r}_{i,k})\times\left(\frac{\vec{r}_{i,k+1}-\vec{r}_{i,k}}{\Delta t}\right) + \left(\frac{\vec{r}_{i,k+1}+\vec{r}_{i,k}}{2}\right)\times(\vec{a}_{i,k}\Delta t)\right]$$

$$= \Delta t\left(\frac{\vec{r}_{i,k+1}+\vec{r}_{i,k}}{2}\right)\times\vec{F}_{i,k}.$$

For notational simplicity, set

$$\vec{T}_{i,k} = \frac{\vec{r}_{i,k+1}+\vec{r}_{i,k}}{2}\times\vec{F}_{i,k},$$

so that

$$\vec{L}_{i,k+1} - \vec{L}_{i,k} = (\Delta t)\vec{T}_{i,k}. \tag{8.45}$$

Hence, if

$$\vec{T}_k = \sum_{i=1}^{3}\vec{T}_{i,k}, \tag{8.46}$$

then (8.43), (8.45), and (8.46) imply

$$\vec{L}_{k+1} - \vec{L}_k = (\Delta t)\vec{T}_k. \tag{8.47}$$

Now, if

$$\vec{T}_k = 0, \quad k = 0,1,2,... \tag{8.48}$$

then

$$\vec{L}_{k+1} = \vec{L}_k, \quad k = 0,1,..., \tag{8.49}$$

which implies (8.44), and the discussion would be complete. It remains for us to show then that for the three-body problem, (8.48) is valid.

To do this observe that

$$\vec{T}_k = \vec{T}_{1,k} + \vec{T}_{2,k} + \vec{T}_{3,k}$$

$$= \frac{\vec{r}_{1,k+1}+\vec{r}_{1,k}}{2}\times\vec{F}_{i,k} + \frac{\vec{r}_{2,k+1}+\vec{r}_{2,k}}{2}\times\vec{F}_{2,k} + \frac{\vec{r}_{3,k+1}+\vec{r}_{3,k}}{2}\times\vec{F}_{3,k}, \tag{8.50}$$

where $\vec{F}_{1,k}$, $\vec{F}_{2,k}$ and $\vec{F}_{3,k}$ are, by (8.12) - (8.14) and the corresponding formu for the y-components:

$$\vec{F}_{1,k} = -\frac{Gm_1m_2(\vec{r}_{1,k+1}+\vec{r}_{1,k}-\vec{r}_{2,k+1}-\vec{r}_{2,k})}{r_{12,k}r_{12,k+1}(r_{12,k}+r_{12,k+1})}$$

$$-\frac{Gm_1m_3(\vec{r}_{1,k+1}+\vec{r}_{1,k}-\vec{r}_{3,k+1}-\vec{r}_{3,k})}{r_{13,k}r_{13,k+1}(r_{13,k}+r_{13,k+1})} \quad (8.51)$$

$$\vec{F}_{2,k} = -\frac{Gm_1m_3(\vec{r}_{2,k+1}+\vec{r}_{2,k}-\vec{r}_{1,k+1}-\vec{r}_{1,k})}{r_{13,k}r_{13,k+1}(r_{13,k}+r_{13,k+1})}$$

$$-\frac{Gm_2m_3(\vec{r}_{2,k+1}+\vec{r}_{2,k}-\vec{r}_{3,k+1}-\vec{r}_{3,k})}{r_{23,k+1}r_{23,k}(r_{23,k+1}+r_{23,k})} \quad (8.52)$$

$$\vec{F}_{3,k} = -\frac{Gm_1m_3(\vec{r}_{3,k+1}+\vec{r}_{3,k}-\vec{r}_{1,k+1}-\vec{r}_{1,k})}{r_{13,k+1}r_{13,k}(r_{13,k}+r_{13,k+1})}$$

$$-\frac{Gm_2m_3(\vec{r}_{3,k+1}+\vec{r}_{3,k}-\vec{r}_{2,k+1}-\vec{r}_{2,k})}{r_{23,k+1}r_{23,k}(r_{23,k}+r_{23,k+1})} \quad (8.53)$$

owever, direct substitution of (8.51) - (8.53) into (8.50) yields, by the laws of ector cross products,

$$\vec{T}_k = \vec{0}, \quad k = 0,1,2,\ldots .$$

hus (8.44), which is called the Law of Conservation of Angular Momentum, is valid.

8.9 EXERCISES - CHAPTER 8

.1 For $\Delta t = 0.01$, $x_{1,0} = 0$, $y_{1,0} = 100$, $x_{2,0} = 100$, $y_{2,0} = 0$, $x_{3,0} = -100$, $y_{3,0} = 0$, and for masses which are normalized so that $m_1 = m_2 = m_3 = 10$, $G = 1$, describe the motion of each particle of the three-body problem up to t_{2000} for each of the following initial velocities:

(a) $v_{1,0,x} = 0$, $v_{1,0,y} = -10$, $v_{2,0,x} = -9$, $v_{2,0,y} = 0$, $v_{3,0,x} = 8$, $v_{3,0,y} = 0$.

(b) $v_{1,0,x} = 0$, $v_{1,0,y} = -10$, $v_{2,0,x} = -10$, $v_{2,0,y} = 0$, $v_{3,0,x} = 0$, $v_{3,0,y} = 0$.

(c) $v_{1,0,x} = -15$, $v_{1,0,y} = 0$, $v_{2,0,x} = -5$, $v_{2,0,y} = 0$, $v_{3,0,x} = 7$, $v_{3,0,y} = -2$.

(d) $v_{1,0,x} = -3$, $v_{1,0,y} = -3$, $v_{2,0,x} = -4$, $v_{2,0,y} = 3$, $v_{3,0,x} = 4$, $v_{3,0,y} = -2$.

8.2 In cgs units with $G = (6.67)10^{-8}$ and for $\Delta t = 0.001$, describe the motion up to t_{5000} of the three-body problem with

$m_1 = 10^8(6.67)^{-1}$, $m_2 = 1$, $m_3 = 0$

$x_{1,0} = 0$, $x_{2,0} = 0.50$, $x_{3,0} = 0$

$y_{1,0} = 0$, $y_{2,0} = 0$, $y_{3,0} = 1000$

$v_{1,0,x} = 0$, $v_{2,0,x} = 0$, $v_{3,0,x} = 0$

$v_{1,0,y} = 0$, $v_{2,0,y} = 1.63$, $v_{3,0,y} = 0$.

Show that, in particular, the motion of P_2 is essentially that of Fig. 7. in Section 7.7.

8.3 In cgs units with $G = (6.67)10^{-8}$ and for $\Delta t = 0.001$, describe the motion up to t_{5000} of the three-body problem with

$m_1 = 10^8(6.67)^{-1}$, $m_2 = 10^6(6.67)^{-1}$, $m_3 = 10^6(6.67)^{-1}$

$x_{1,0} = 0$, $x_{2,0} = 0.50$, $x_{3,0} = -1$

$y_{1,0} = 0$, $y_{2,0} = 0$, $y_{3,0} = 8$

$v_{1,0,x} = 0$, $v_{2,0,x} = 0$, $v_{3,0,x} = 0$

$v_{1,0,y} = 0$, $v_{2,0,y} = 1.63$, $v_{3,0,y} = -3.75$.

In particular, compare this motion with those of the examples in Section 8.

8.4 In a plane, simulate the motion of a Sun-Earth-Jupiter system.

8.5 In a plane, simulate the motion of a Sun-Earth-Moon system.

8.6 Show how to determine the center of gravity of any plane figure by hanging it from two different points of its boundary.

8.7 Find the center of gravity at time t_0 of the three-body systems of Exercises 8.1, 8.2 and 8.3.

8.8 Show directly from your computations that the motion of the center of gravi for each of Exercises 8.1, 8.2 and 8.3 is along a straight line.

8.9 Show directly from your computations that the linear momentum of each syste is conserved for the motions of Exercises 8.1 - 8.5.

8.10 Show directly from your computations that the angular momentum of each syst is conserved for the motions of Exercises 8.1 - 8.5.

8.11 Formulate the three-body problem in three dimensions and prove the laws of conservation of energy and of linear momentum. What is the nature of the motion of the center of gravity?

CHAPTER 9

The n-Body Problem

9.1 INTRODUCTION

this chapter we will continue in the essential spirit of the last chapter and rmulate an approach to the general n-body problem which conserves energy, linear mentum and angular momentum. However, rather than repeating all the proofs given r the three-body problem and showing how readily these extend to an arbitrary mber of interacting bodies, we will, instead, explore some new models of two other portant physical phenomena, namely, heat conduction and elasticity. These studies ll require some intuition about the local interaction of very "small particles", ke molecules, in solids. This intuition is very simple, indeed, for when small rticles are "far apart" they attract, when they are "close together" they repel, d the repulsion is much stronger than is the attraction. Thus, instead of the $/r^2$ force of attraction," we might expect a force with a "$1/r^6$ component of traction" and a "$1/r^9$ component of repulsion." Such forces were alluded to ready in Section 7.9.

9.2 DISCRETE n-BODY INTERACTION

gain, for positive time step Δt, let $t_k = k\Delta t$, $k = 0,1,2,\ldots$. At time t_k t particle P_i of mass m_i be located at $\vec{r}_{i,k} = (x_{i,k}, y_{i,k})$, have velocity $_{,k} = (v_{i,k,x}, v_{i,k,y})$, and have acceleration $\vec{a}_{i,k} = (a_{i,k,x}, a_{i,k,y})$, for $i = 1,$ $\ldots,n$. Position, velocity, and acceleration are assumed to be related by

$$\frac{\vec{v}_{i,k+1} + \vec{v}_{i,k}}{2} = \frac{\vec{r}_{i,k+1} - \vec{r}_{i,k}}{\Delta t} \tag{9.1}$$

$$\vec{a}_{i,k} = \frac{\vec{v}_{i,k+1} - \vec{v}_{i,k}}{\Delta t}. \tag{9.2}$$

$\vec{F}_{i,k} = (F_{i,k,x}, F_{i,k,y})$ is the force acting on P_i at time t_k, then force d acceleration are assumed to be related by the discrete dynamical equation

$$\vec{F}_{i,k} = m_i \vec{a}_{i,k}.$$ (9.3

In particular, we now choose $\vec{F}_{i,k}$ to have a component of attraction which beha like G/r^{α} and a component of repulsion which behaves like H/r^{β}, where G, H, and β are nonnegative parameters which are fixed in any particular problem with $\alpha \geq 2$, $\beta > 2$, and where r is the distance between a given pair of particles. For this purpose, let $r_{ij,k}$ be the distance between P_i and P_j at t_k. The $\vec{F}_{i,k}$, the force exerted on P_i by the remaining particles, is defined, in anal with (7.47) and (7.48), by

$$\vec{F}_{i,k} = m_i \sum_{\substack{j=1 \\ j \neq i}}^{n} \left\{ m_j \left(- \frac{G \sum_{\xi=0}^{\alpha-2} (r_{ij,k}^{\xi} r_{ij,k+1}^{\alpha-\xi-2})}{r_{ij,k}^{\alpha-1} r_{ij,k+1}^{\alpha-1} (r_{ij,k+1}+r_{ij,k})} + \frac{H \sum_{\xi=0}^{\beta-2} (r_{ij,k}^{\xi} r_{ij,k+1}^{\beta-\xi-2})}{r_{ij,k}^{\beta-1} r_{ij,k+1}^{\beta-1} (r_{ij,k+1}+r_{ij,k})} \right) \times (\vec{r}_{i,k+1}+\vec{r}_{i,k}-\vec{r}_{j,k+1}-\vec{r}_{j,k}) \right\}, \quad i = 1,2,\ldots,n.$$ (9.4

The summation in (9.4) implies that the forces due to all particles different fr P_i are being combined to determine the force on P_i. Also, formula (9.4) is so general that it includes both gravitational and molecular type particle forces, can be seen, by setting $\alpha = 2$, $n = 3$, $H = 0$ and $G = (6.67)10^{-8}$.

9.3 THE SOLID STATE BUILDING BLOCK

Let us consider next developing a viable model of a solid. In doing this, we wi attempt to simulate contemporary physical thought, in which moleclues and atoms exhibit small vibrations within the solid. Hence, consider first a system of on two particles, P_1 and P_2, of equal mass, which interact according to (9.4). Assume that the force between the particles is <u>zero</u>. Then, from (9.4),

$$\frac{-G \sum_{\xi=0}^{\alpha-2} (r_{ij,k}^{\xi} r_{ij,k+1}^{\alpha-\xi-2})}{r_{ij,k}^{\alpha-1} r_{ij,k+1}^{\alpha-1} (r_{ij,k+1}+r_{ij,k})} + \frac{H \sum_{\xi=0}^{\beta-2} (r_{ij,k}^{\xi} r_{ij,k+1}^{\beta-\xi-2})}{r_{ij,k}^{\beta-1} r_{ij,k+1}^{\beta-1} (r_{ij,k+1}+r_{ij,k})} = 0.$$ (9.5

But, if there is zero force between the two particles, then $r_{ij,k} = r_{ij,k+1}$, s set $r_{ij,k} = r_{ij,k+1} = r$ in (6.5) to yield

$$\frac{-G \sum_{\xi=0}^{\alpha-2} r^{\alpha-2}}{r^{2\alpha-2}} + \frac{H \sum_{\xi=0}^{\beta-2} r^{\beta-2}}{r^{2\beta-2}} = 0.$$ (9.6

hus, for $\beta > \alpha$, which is physically reasonable,

$$- Gr^{-\alpha}(\alpha-1) + Hr^{-\beta}(\beta-1) = 0,$$

r, finally,

$$r = \left[\frac{H(\beta-1)}{G(\alpha-1)}\right]^{1/(\beta-\alpha)}, \quad \beta > \alpha \geq 2. \tag{9.7}$$

onsider next a system of only three particles, P_1, P_2 and P_3, of equal masses, nd assume that no force acts between any two of the particles. Then the distance etween any two of the particles is given, again, by (9.7). Such a configuration s therefore exceptionally stable and will be called a triangular building block.

hen considering a solid we will decompose it into triangular building blocks. hen, by an appropriate choice of parameters, the force on any particle of a trianular block due to more distant particles will be made small, thus achieving the mall vibrations desired. To illustrate, let the six particles P_1, P_2, P_3, P_4, $_5$, P_6 be located at the vertices of the four triangular building blocks of the riangular region OAB, shown in Fig. 9.1. Assume that $m_i \equiv 1$, $G = H = 1$, $= 7$, and $\beta = 10$, so that $r = \sqrt[3]{1.5}$. The particles' initial positions are, hen,

$$(x_1,y_1) = (1.14471,\ 1.98270), \quad (x_2,y_2) = (0.57236,\ 0.99135)$$

$$(x_3,y_3) = (1.71707,\ 0.99135), \quad (x_4,y_4) = (0,0)$$

$$(x_5,y_5) = (1.14471,\ 0), \quad (x_6,y_6) = (2.28943,\ 0).$$

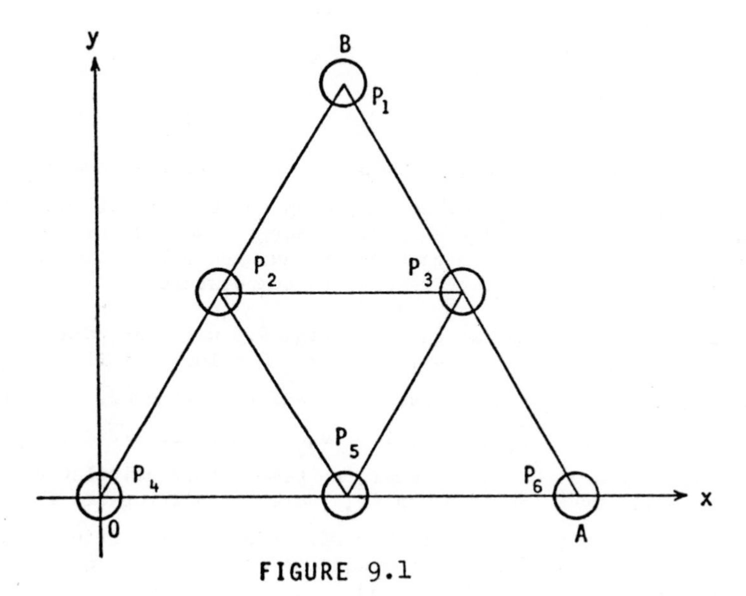

FIGURE 9.1

Assign to each particle a $\vec{0}$ initial velocity. Finally, let particles P_4 and P_6 be fixed by defining the total force on each of these particles to be $\vec{0}$ and allow the remaining particles to move under force law (9.4). For $\Delta t = 0.05$ and for 2500 time steps, the motions of P_1, P_2, P_3 and P_5 were generated from (9.1) - (9.4). P_1 and P_5 exhibited small oscillations in the vertical directic only, while P_2 and P_3 exhibited small two dimensional oscillations. The max mum distance, for example, that P_1 moved from its initial position was approximately 0.02, and this occurred at approximately every one hundred time steps. The running time on the UNIVAC 1108 was 4 minutes and a general n-body FORTRAN is given in Appendix E.

9.4 FLOW OF HEAT IN A BAR

Let us now develop the basic concepts of discrete conductive heat transfer by con centrating on the prototype problem of heat flow in a bar. Physically, the probl is formulated as follows. Let the region bounded by rectangle OABC, as shown i Fig. 9.2, represent a bar. Let $|OA| = a$, $|OC| = c$. A section of the boundary the bar is heated. The problem is to describe the flow of heat through a bar.

Our discrete approach to the problem proceeds as follows. First, subdivide the given region into triangular building blocks, one such possible subdivision of which is shown in Fig. 9.3 for the parameter choices $m_i \equiv 1$, $G = H = 1$, $\alpha = 7$, $\beta = 10$, $a = 11$, $c = 2$. In this case, from (9.7), $r \sim 1.1447142426$.

Now, by <u>heating</u> a section of the boundary of the bar, we will mean <u>increasing the velocity</u>, and hence the kinetic energy, of some of the particles whose centers ar on OABC. By the <u>temperature</u> $T_{i,k}$ of particle P_i at time t_k, we will mean the following. Let M be a fixed positive integer and let $K_{i,k}$ be the kinetic energy of P_i at t_k. Then $T_{i,k}$ is defined by

$$T_{i,k} = \frac{1}{M} \sum_{j=k-M+1}^{k} K_{i,j},$$

which is, of course, the arithmetic mean of P_i's kinetic energies at M consec tive time steps. By the <u>flow</u> of heat through the bar we will mean the transfer t other particles of the bar of the kinetic energy added at the boundary. Finally, to follow the flow of heat through the bar one need only follow the motion of eac particle and, at each time step, record its temperature.

To illustrate, consider the bar shown in Fig. 9.3 with the parameter choices give above, that is, $m_i \equiv 1$, $G = H = 1$, $\alpha = 7$, $\beta = 10$, $a = 11$, $c = 2$. Assume th a strong heat source is placed above P_6, and then removed, in such a fashion th $\vec{v}_{5,0} = (-\sqrt{2}/2, -\sqrt{2}/2)$, $\vec{v}_{6,0} = (0,-1)$, $\vec{v}_{7,0} = (\sqrt{2}/2, -\sqrt{2}/2)$, while all other initial velocities are $\vec{0}$. With regard to temperature calculation, assume that t velocities of all particles prior to t_0 were $\vec{0}$. As regards the choice of M, which is completely arbitrary, set $M = 20$. From the resulting calculations with $\Delta t = 0.025$, Figures 9.4 - 9.8 show the constant temperature contours $T = 0.1$, 0.0 0.025, 0.002 at t_5, t_{10}, t_{15}, t_{20} and t_{25}, respectively. The resulting wave motion is clear and Fig. 9.8 exhibits wave reflection.

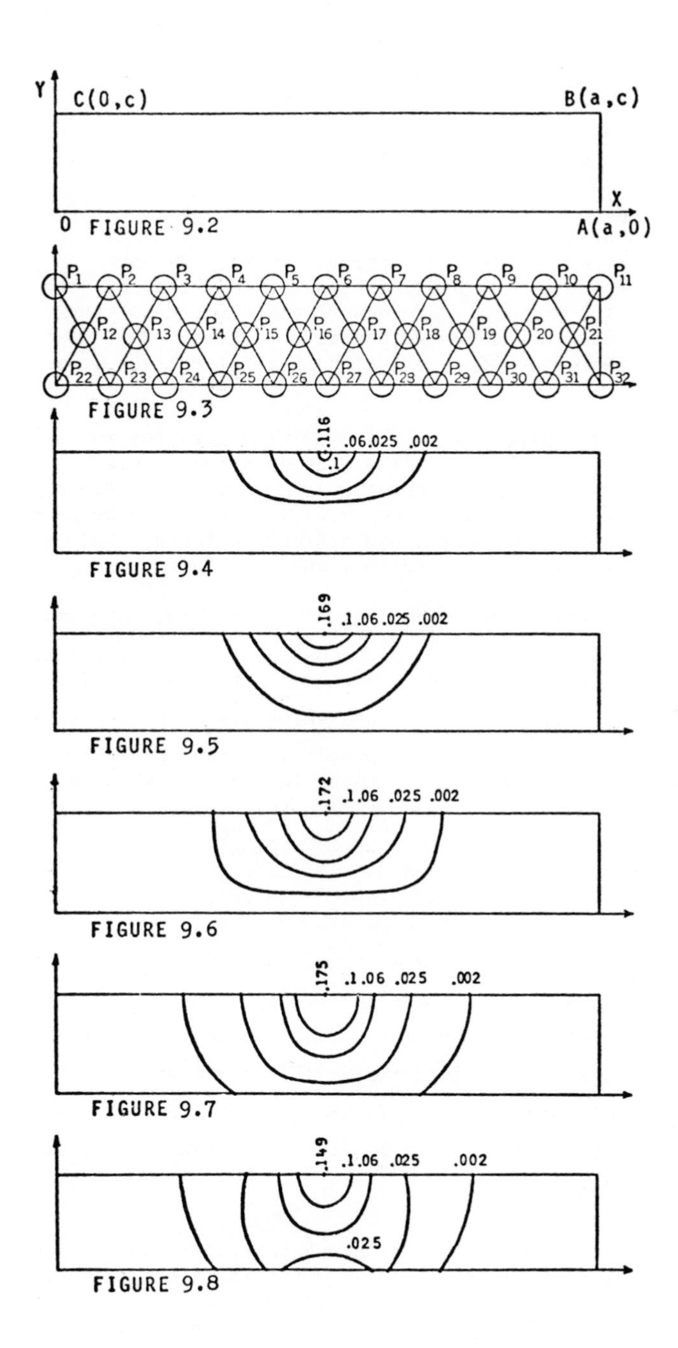

FIGURE 9.2

FIGURE 9.3

FIGURE 9.4

FIGURE 9.5

FIGURE 9.6

FIGURE 9.7

FIGURE 9.8

Other heat transfer concepts can be defined now in the same spirit as above, as follows. A side of a bar is insulated means that the bar particles cannot transf energy across this side of the bar to particles outside the bar, as in the above example, while melting is the result of adding so much heat that various particle attain sufficiently high speeds which break the bonding effect of (9.4).

9.5 OSCILLATION OF AN ELASTIC BAR

Next, let us develop the basic mechanisms of discrete elasticity by concentrating on the vibration of an elastic bar. The problem is formulated physically as follows. Let the region bounded by rectangle OABC, as shown in Fig. 9.2, represen a bar which can be deformed, and which, after deformation, tends to return to its original shape. The problem is to describe the motion of such a bar after an external force, which has deformed the bar, is removed. Equivalently, the proble is to describe the motion of an elastic bar after release from a position of tension.

Our discrete approach proceeds as follows. The given region is first subdivided into triangular building blocks. Then, deformation results in the compression of certain paraticles and the stretching apart of others. Release from a position o deformation, or tension, results, by (9.4), in repulsion between each pair of par ticles which have been compressed and attraction between each pair which have bee stretched, the net effect being the motion of the bar.

As a particular example, let $m_i \equiv 1$, $\alpha = 7$, $\beta = 10$, $G = 425$, $H = 1000$, and $\Delta t = .025$. From (9.7), $r = 1.52254$. Consider, for variety, the thirty particle bar which results by deleting P_{11} and P_{32} from the configuration of Fig. 9.3. The particles P_1, P_{12}, and P_{22}, whose respective coordinates are (0, 2.63711), (.76127, 1.31855), and (0,0), are to be held fixed throughout. This is done by defining the total force on each of these two particles to be $\vec{0}$. In order to obtain an initial position of tension like that shown in Fig. 9.14a, first set P_{13}, P_{14}, P_{15}, P_{16}, P_{17}, P_{18}, P_{19}, P_{20} and P_{21} at (2.28357, 1.29198), (3.80588, 1.26541), (5.32632, 1.18573), (6.84052, 1.02658), (8.33992, .76219), (9.81058, .36813), (11.23199, -.17750), (12.57631, -.89228), and (13.80807, -1.78721), respectively. Any two consecutive points P_k, P_{k+1}, $k = 13,14,\ldots,20$, are pos tioned r units apart. The points $P_2 - P_{10}$ and $P_{23} - P_{31}$ are then positione as follows: P_{k-10} and P_{k+11} are the two points which are r units from both P_k and P_{k+1} for each of $k = 12,13,\ldots,20$. Each consecutive pair of points in the $P_2 - P_{10}$ set are then separated by a distance greater than r, while each consecutive pair of points in the $P_{23} - P_{31}$ set are separated by a distance les than r. Thus, the points $P_2 - P_{10}$ are in a stretched position, while the poin $P_{23} - P_{31}$ are compressed.

From the initial position of tension shown in Fig. 9.14a, the oscillatory motion the bar is determined from (9.1) - (9.4) with all initial velocities set at $\vec{0}$. T upward swing of the bar was plotted automatically at every twenty time steps and shown in Fig. 9.14a-l from t_0 to t_{220}. It is of interest to note that as the bar moves, each row of particles exhibits wave oscillation and reflection.

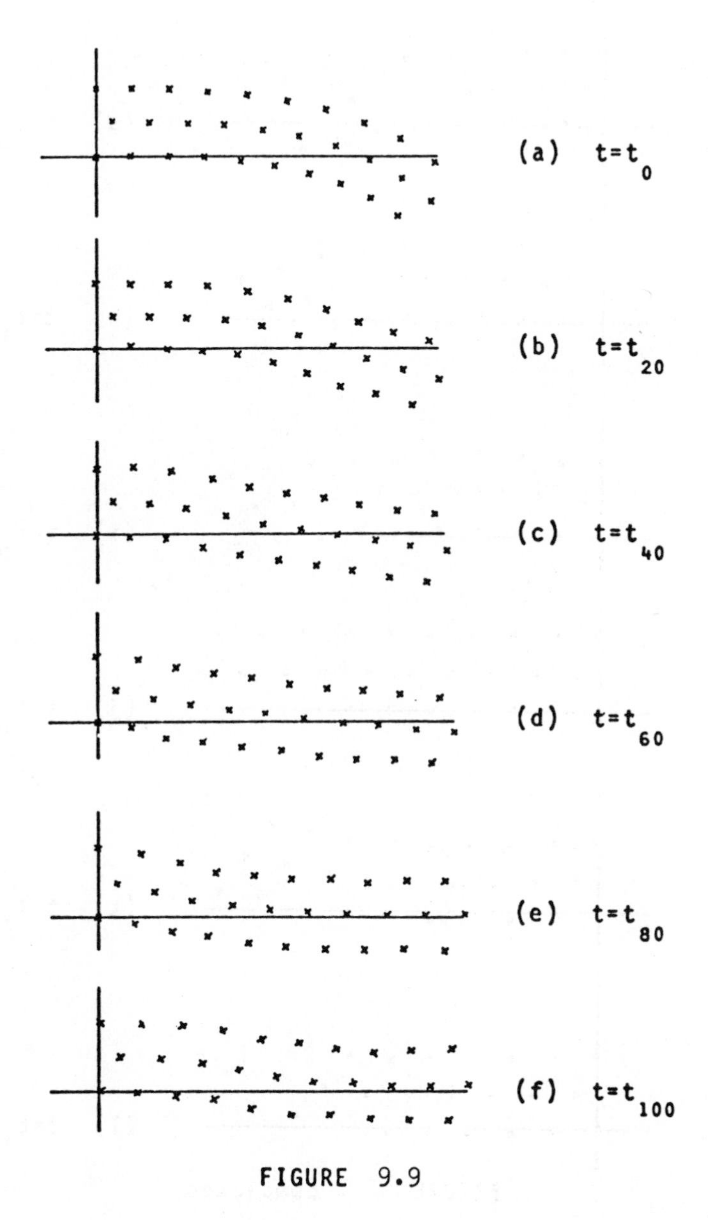

FIGURE 9.9

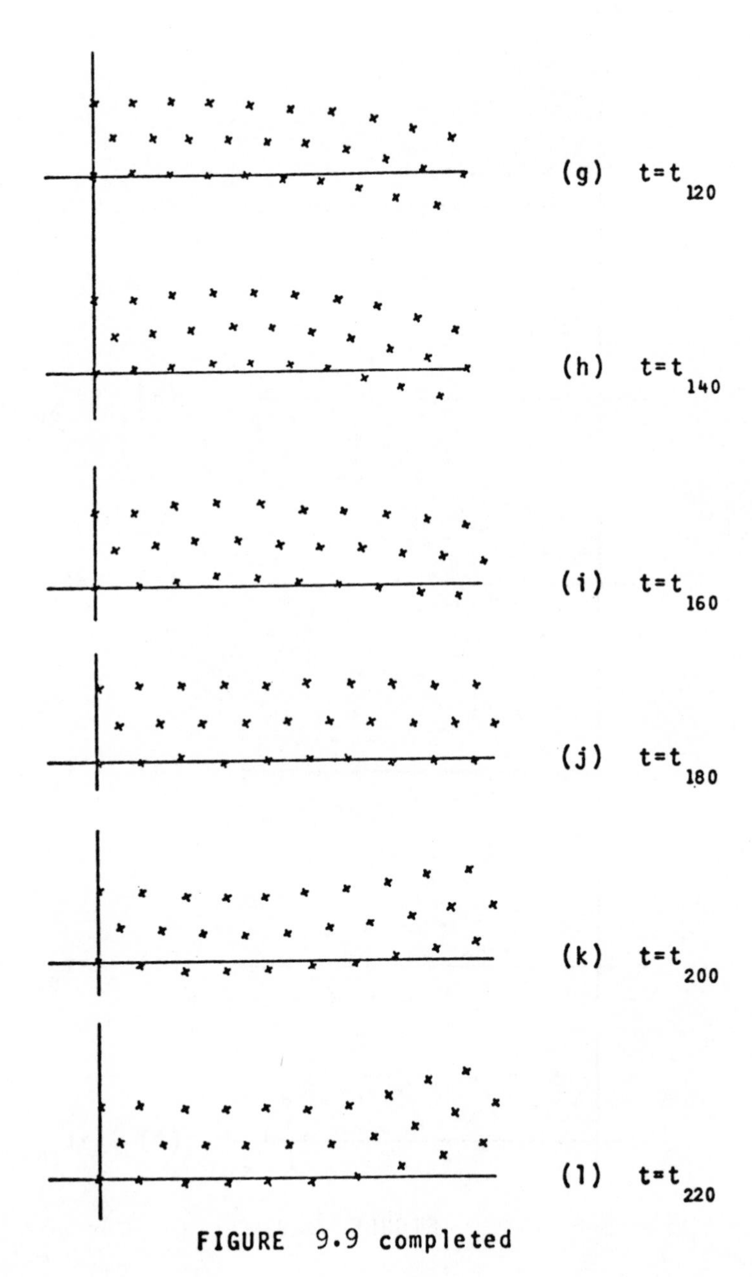

FIGURE 9.9 completed

9.6 EXERCISES - CHAPTER 9

.1 Write out the force components of (9.4) which act on each particle for the following sets of parameter values.

(a) $n = 2$, $\alpha = 2$, $\beta = 2$, $G = (6.67)10^{-8}$, $H = 0$.

(b) $n = 3$, $\alpha = 2$, $\beta = 2$, $G = (6.67)10^{-8}$, $H = 0$.

(c) $n = 2$, $\alpha = 6$, $\beta = 9$, $G = 1$, $H = 1$.

(d) $n = 3$, $\alpha = 6$, $\beta = 12$, $G = 1$, $H = 1$.

(e) $n = 10$, $\alpha = 2$, $\beta = 2$, $G = (6.67)10^{-8}$, $H = 0$.

.2 Determine the length of the side of a triangular building block for each of the following parameter choices.

(a) $\alpha = 2$, $\beta = 3$, $G = H = 1$.

(b) $\alpha = 2$, $\beta = 4$, $G = H = 1$.

(c) $\alpha = 6$, $\beta = 9$, $G = H = 1$.

(d) $\alpha = 6$, $\beta = 12$, $G = H = 1.2$.

(e) $\alpha = 6$, $\beta = 9$, $G = 1.0$, $H = 1.2$.

.3 Describe the flow of heat in the conduction example of Section 9.4 if the parameter change $M = 1$ is made, so that a particle's temperature is identified with its kinetic energy. Compare the two choices $M = 1$ and $M = 20$.

.4 Describe the flow of heat in the conduction example of Section 9.4 if the heat source is not removed.

.5 Describe the flow of heat in the conduction example of Section 9.4 if the side OA is not insulated.

.6 Describe the flow of heat in the conduction example of Section 9.4 if the source of heat is applied only at the end AB.

.7 Give an example of the flow of heat in a three dimensional bar.

.8 Describe the oscillatory motion of the example of Section 9.5 from t_{220} to t_{500}.

.9 Give an example of the oscillatory motion from a position of tension of a three dimensional elastic bar.

.10 Simulate the elastic buckling of half a tennis ball which "pops" inside out under pressure on its outer surface.

.11 Assuming that all the planets rotate in the same plane, simulate ten years of solar system motion.

.12 Develop a model of friction and run a computer example to test its viability.

.13 Prove that (9.1) - (9.4) imply the conservation of energy.

CHAPTER 10

Discrete Fluid Models

10.1 INTRODUCTION

Intuitively, anything which is either a gas or a liquid is called a fluid. Thus, oxygen, nitrogen, ozone, air, water, oil, and blood are all fluids. The number o phenomena which involve fluids is so great that we could study these alone and nev want for unanswered questions. For example, fluid problems are basic to the studi of weather development, ocean flow, aerodynamics, animal circulatory systems, and living cell metabolism.

In this chapter we will study some basic, interesting phenomena in fluid flow. Simultaneously, we will explore and evaluate the viability of both conservative a nonconservative models for these problems.

10.2 LAMINAR AND TURBULENT FLOWS

Let us examine the flow of a fluid first as a conservative, n-body model, using t definitions and formulas of Chapter 9. The only possible change we might want to make is that (9.4) be adjusted to include the effects of gravity. Thus, in place of (9.4), we could use, in, say, lb.-foot-sec. units,

$$\vec{\mathfrak{F}}_{i,k} = \vec{F}_{i,k} - 32\vec{\gamma}m_i, \tag{10.1}$$

where $\vec{\gamma} = (0,1)$ in two dimensions, or $\vec{\gamma} = (0,0,1)$ in three dimensions. Howeve as indicated, the scope of fluid dynamics is so vast that we are forced to concen trate here only on some small portion of it. Hence, for simplicity, and because the resulting flows will be relatively dramatic in their behavior, we will concen trate on relatively high-speed fluid interactions in which the effects of gravity will be negligible. A broader scope of study would require replacement of (9.4) (10.1) and localization and rescaling of molecular type forces (Greenspan (1980))

Consider then a two dimensional liquid in motion, a small portion of which is sho in Fig. 10.1. Let particles $P_1 - P_{11}$ be called the first row, $P_{12} - P_{23}$ the second row, and $P_{24} - P_{34}$ the third row. In (9.4), let $G \equiv H \equiv m_i \equiv 1$, $i = 1$ $2,\ldots,34$, and $\alpha = 7$, $\beta = 10$. The initial positions of $P_1 - P_{34}$ are set so that $P_{13} - P_{22}$ are centers of regular hexagons of radii $r = (1.5)^{1/3}$, while t

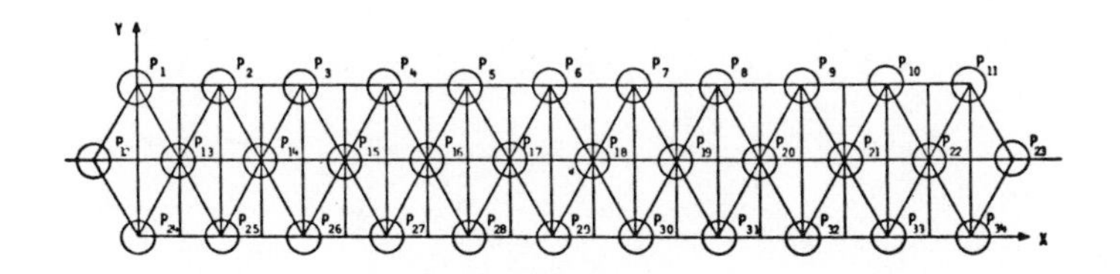

FIGURE 10.1

maining particles are centered at the vertices of the hexagons. This choice of is, of course, by (9.7), one of relative configuration stability.

e motion of the particles will be completely determined, as described in Chapter once we fix all the initial velocities $v_{i,0,x}$ and $v_{i,0,y}$. To do this, let us ppose that the particles have just been emitted horizontally from a nozzle. If is were the case, then $v_{i,0,x}$ would dominate $v_{i,0,y}$. Moreover, not all parties would have exactly the same velocities because of possible collisions with the zzle housing, and so forth. So, let us choose

$$v_{i,0,x} = V + \varepsilon_{i,1}, \quad v_{i,0,y} = \varepsilon_{i,2}, \quad i = 1,2,\ldots,34 \tag{10.2}$$

ere V is a parameter which assures relatively horizontal motion, while $\varepsilon_{i,1}$ d $\varepsilon_{i,2}$ are relatively small random numbers which give the particles small perrbations from purely horizontal motion. For simplicity, let the computer genere all the $\varepsilon_{i,1}$ and $\varepsilon_{i,2}$ in a random fashion so that

$$|\varepsilon_{i,j}| \le (1\%)V = \frac{V}{100}, \quad j = 1,2. \tag{10.3}$$

the examples which follows, our interest will center on increasing values of V d on initial time steps only. The random numbers are generated by the computer ly once and independently of V. However, in each example, these numbers are scaled proportionately so that the maximum $|\varepsilon_{i,j}|$ always gives equality in 0.3).

gure 10.2 shows the particle motion for $V = 50$ and $\Delta t = 0.02$ at $t = 0.2$, 4, 0.6, 0.8, 1.0, 1.2 and 1.4. A gentle wave motion develops in each row while e rows maintain their relative positions. A flow of this nature is said to be minar. Figure 10.3 shows the motion for $V = 300$ and $\Delta t = 0.02$ at $t = 0.2$, 4, 0.6, 0.8 and 1.0. Repulsion between the particles has assumed a greater sigficance and, though the rows still maintain their relative positions, the motion becoming more chaotic. Figure 10.4 shows the motion for $V = 1000$ with $= 0.1$ at $t = 0.2$, 0.4, 0.6, 0.8 and 1.0. So much motion results that the oice $\Delta t = 0.1$ was necessary for the convergence of Newton's method. Here, the minar character of the flow has disappeared in that the rows no longer maintain eir relative positions, and the motion becomes relatively chaotic, or, more scriptively, turbulent. Thus, with the increase in velocity, particles can come arer to other particles, which results in increased repulsive forces and more mplex motion.

tuitively, turbulence is often thought of as a type of fluid flow which is charterized by the rapid appearance and disappearance of many small vortices, or

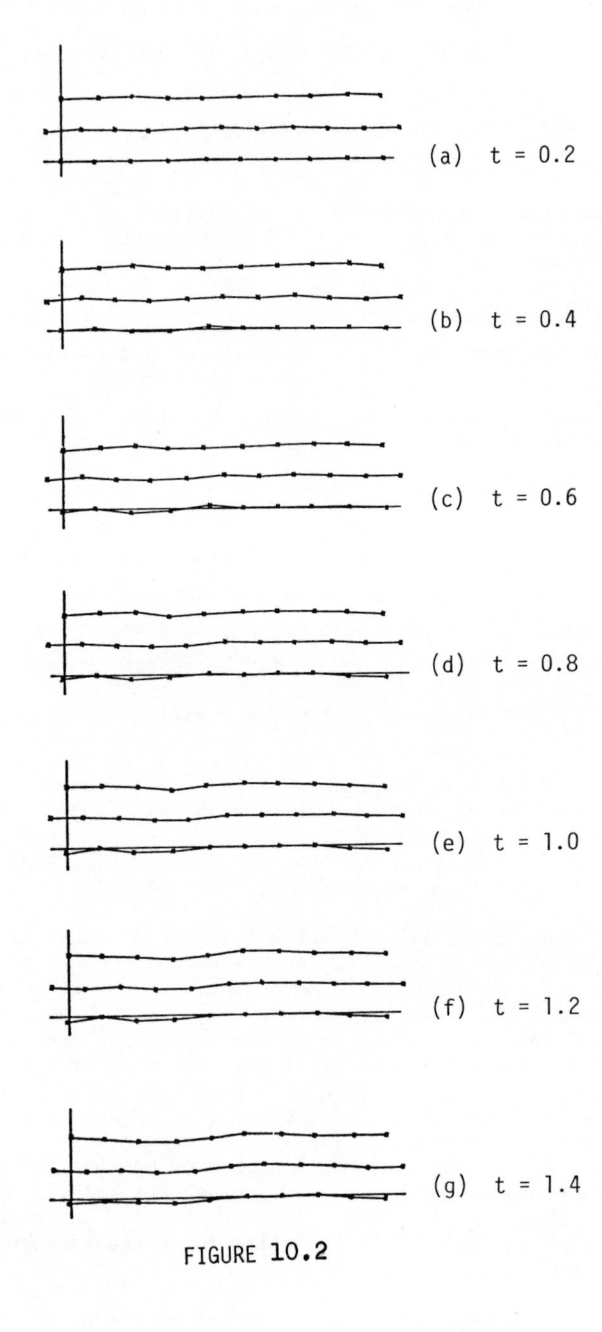

FIGURE 10.2

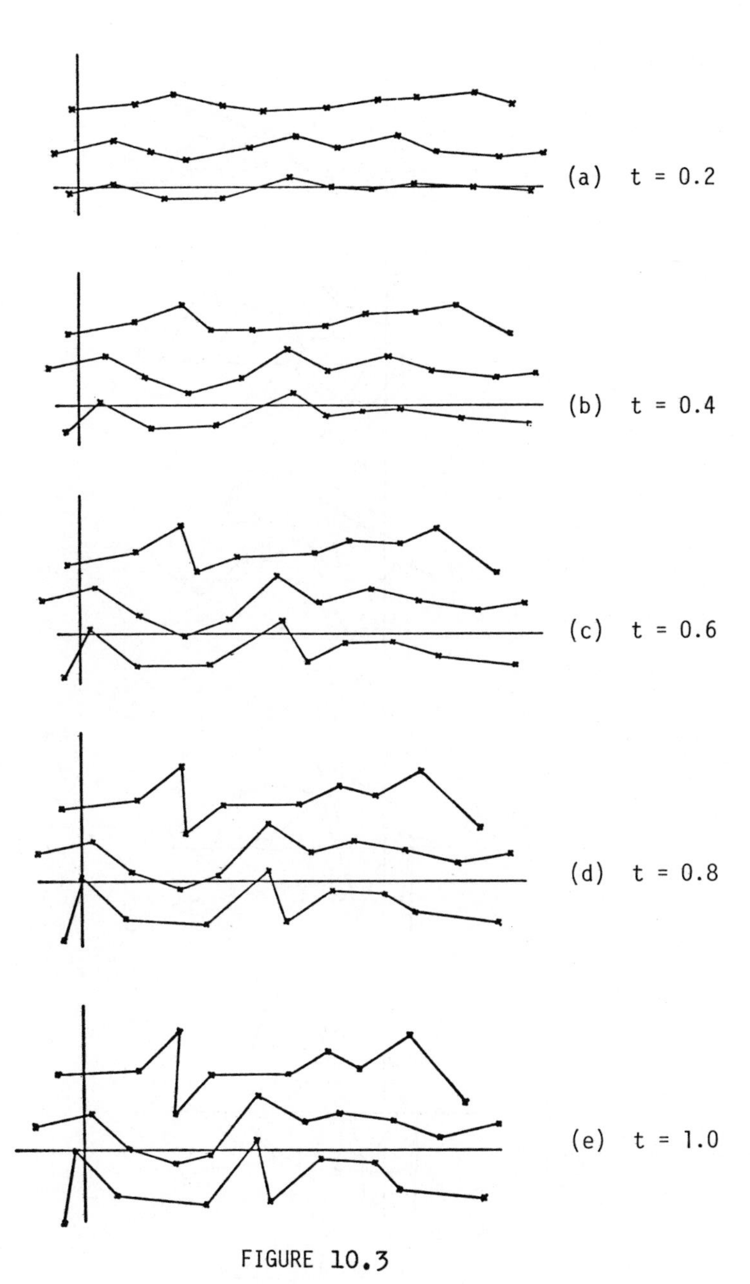

FIGURE 10.3

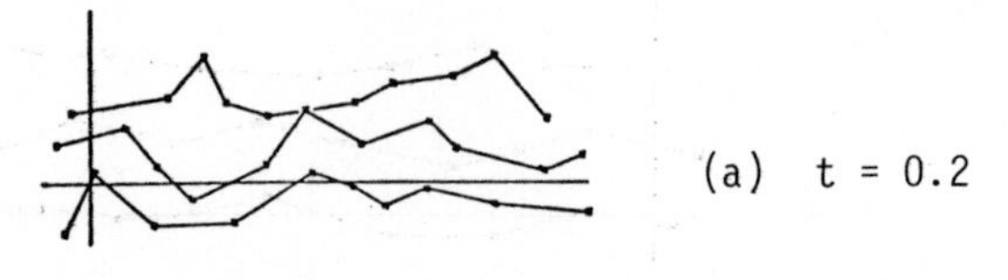

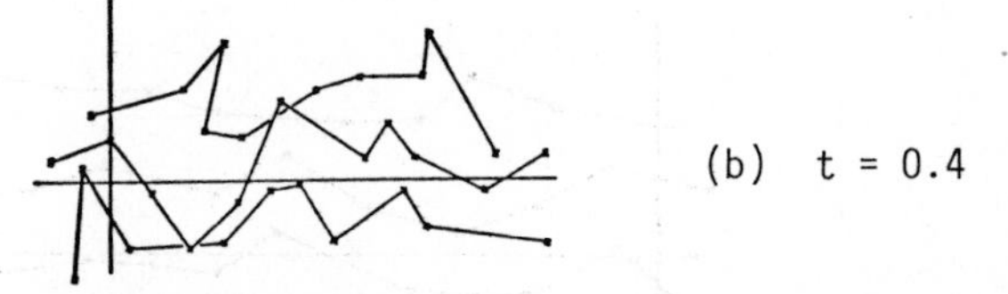

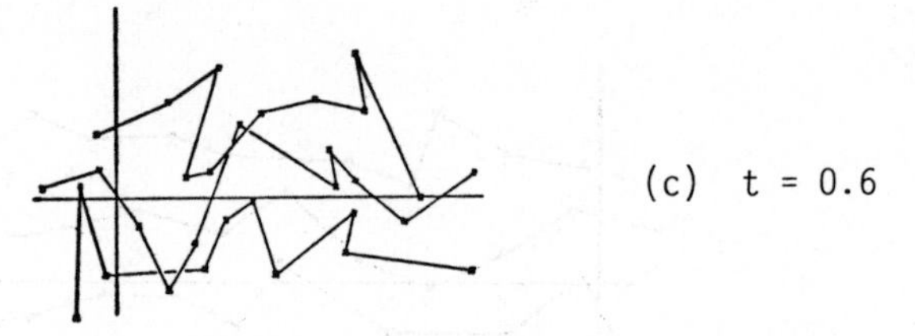

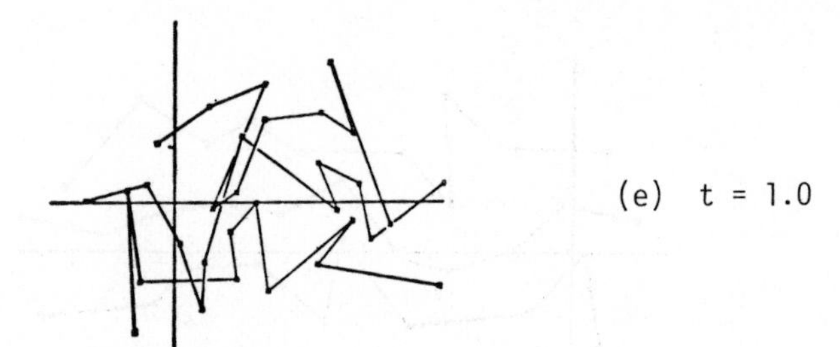

FIGURE 10.4

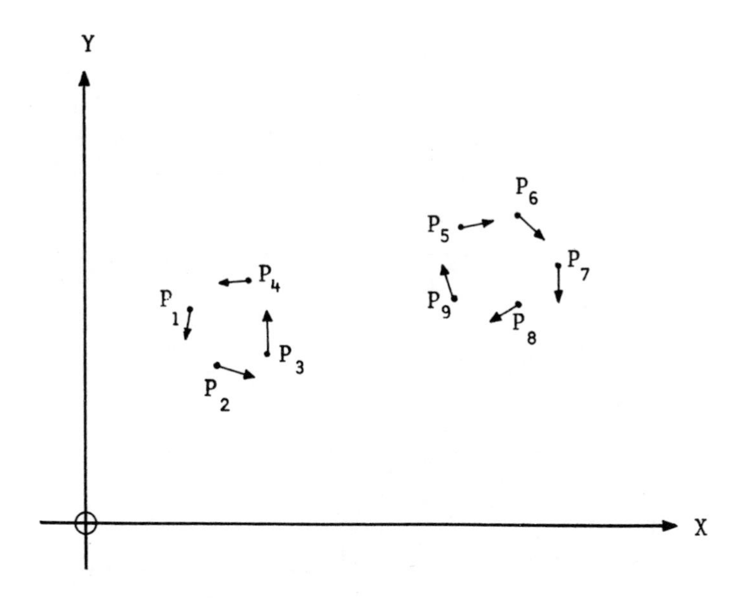

FIGURE 10.5

hirlpools. If one defines a vortex as a counterclockwise, or a clockwise, motion xhibited by three or more relatively close particles, as shown in Fig. 10.5, then, ndeed, such configurations do appear in our example for $V = 1000$, do break down uickly due to the very large effects of repulsion, and then do reappear in differ- nt particle groupings.

he importance of the concept of turbulence cannot be overestimated, since it is he most common, yet least understood, type of fluid flow in both nature and in echanical devices (see, e.g., von Karman).

10.3 SHOCK WAVES

n Section 10.2, we saw how easy it was to model the change of a fluid's flow from hat which was laminar to that which was turbulent. Though conceptually simple, owever, that model is expensive practically. The reason is that at each time step e must solve a large number of equations, or, to be precise, if there are n par- icles, then, at each time step, a two dimensional model requires the solution of n equations and a three dimensional model requires the solution of $6n$ equations. f $n = 10^5$, the problem becomes voluminous from the point of view of existing omputer technology. For such problems, then, we forsake the conservation proper- ies and, in the sense of Definition 2.4, turn from conservative, implicit models o nonconservative, explicit models. To illustrate, let us examine the very inter- sting concept of a shock wave.

n contrast with a liquid, a gas has relatively few particles per unit of volume. onsider, then, a gas as shown in a long tube in Fig. 10.6(a). Into this tube nsert a piston, as shown in Fig. 10.6(b). If one first moves the piston down the ube slowly, then, as shown in Fig. 10.6(c), the gas particles increase in density er unit volume in a relatively uniform way. However, if, as shown in Fig. 10.6(d), he piston is moved at a very high rate of speed, then gas particles compact on the

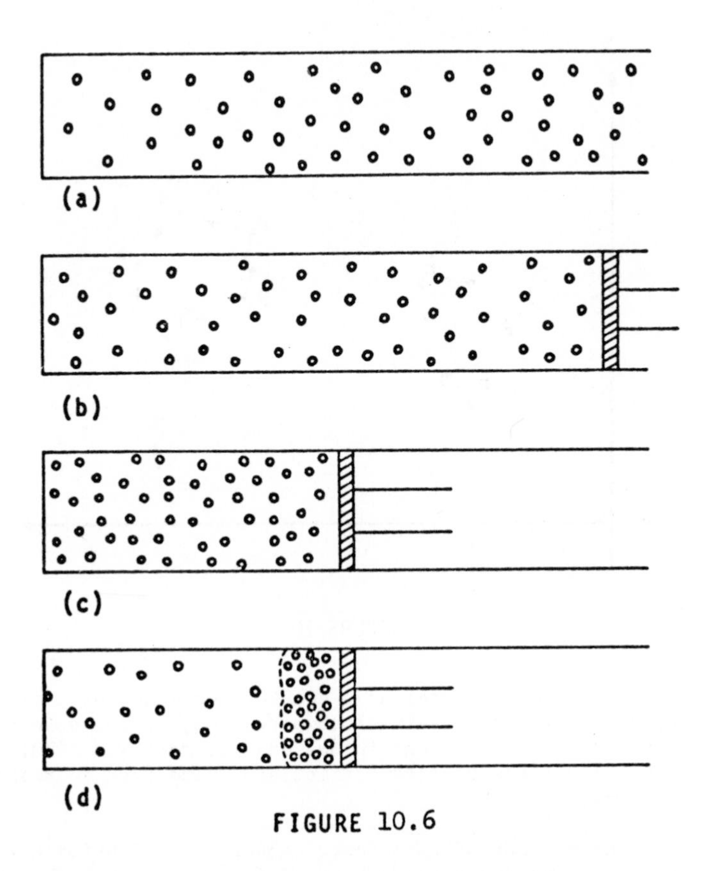

FIGURE 10.6

cyclinder head, with the result that the original gas consists of two distinct po tions, one with a very high density, the other with about the same density as at the start. The boundary between these two portions, which is shown as a dotted line in Fig. 10.6(d), is called a shock wave.

More generally, a shock wave can be thought of as follows. Assign to a given gas positive measure of average particle density. Let a body B pass through the ga at a very high rate of speed. In certain regions about B, there may occur sets of gas particles whose densities are not average. Then, a boundary between sets particles with average density and those with "greater than average" density is called a shock wave.

Let us illustrate this "greater than average" density concept and the development of a shock wave by considering next a particular shock tube problem. Consider th tube configuration in Fig. 10.6(b). For convenience, a coordinate system will be fixed relative to the piston head, as shown in Fig. 10.7, so that the particles will be considered to be in motion relative to the piston. Let the tube be 100 units long, so that $AO = 100$, and 10 units high, so that $AB = 10$. Now, ever $\Delta t = 0.01$ seconds, let a column of particles, each of radius $r = 0.35$ and of un mass m, enter the tube at AB. Each such column is determined as follows. At each time t_k, each position $(-100, n+\frac{1}{2})$, $n = 0,1,2,\ldots,9$, is either filled b a particle or left vacant by a random process, like the toss of a coin. Once it has been determined that a particle P_i is at such an initial location, then its initial velocity $\vec{v}_{i,0}$ is determined by

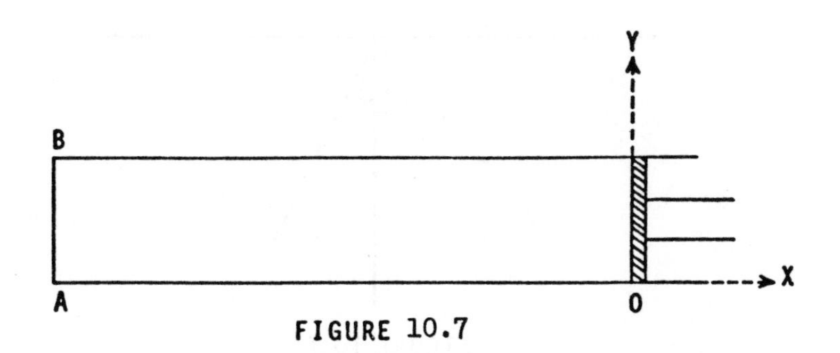

FIGURE 10.7

$$\vec{v}_{i,0} = (100+\varepsilon_{i,1}, \varepsilon_{i,2}),$$

where $\varepsilon_{i,1}$ and $\varepsilon_{i,2}$ are selected at random, but are small in magnitude relative to 100, thus assuring that the gas has a relatively high speed in a relatively uniform direction. Neglecting attractive forces, but allowing a repulsive force to simulate the effects of particle collision, we will take an n-body formulation in which the position $(x_{i,k}, y_{i,k})$ of P_i at time t_k is given explicitly by

$$v_{i,k+1,x} = v_{i,k,x} + a_{i,k,x}\Delta t \tag{10.4}$$

$$v_{i,k+1,y} = v_{i,k,y} + a_{i,k,y}\Delta t \tag{10.5}$$

$$x_{i,k+1} = x_{i,k} + \frac{\Delta t}{2}(v_{i,k+1,x}+v_{i,k,x}) \tag{10.6}$$

$$y_{i,k+1} = y_{i,k} + \frac{\Delta t}{2}(v_{i,k+1,y}+v_{i,k,y}) \tag{10.7}$$

$$a_{i,k,x} = \frac{1}{m}\sum_{\substack{j=1\\ j\neq i}}^{n} \frac{(x_{i,k}-x_{j,k})H}{(r_{ij,k}+\xi)^{p+1}} \tag{10.8}$$

$$a_{i,k,y} = \frac{1}{m}\sum_{\substack{j=1\\ j\neq i}}^{n} \frac{(y_{i,k}-y_{j,k})H}{(r_{ij,k}+\xi)^{p+1}} \tag{10.9}$$

where

$$r_{ij,k} = \text{distance between } P_i \text{ and } P_j, \text{ in the tube, at } t_k \tag{10.10}$$

$$n = \text{total number of particles in the tube at } t_k \tag{10.11}$$

$$H = \begin{cases} 0, & \text{if } r_{ij,k} \geq 2r \\ 1, & \text{if } r_{ij,k} < 2r, \end{cases} \tag{10.12}$$

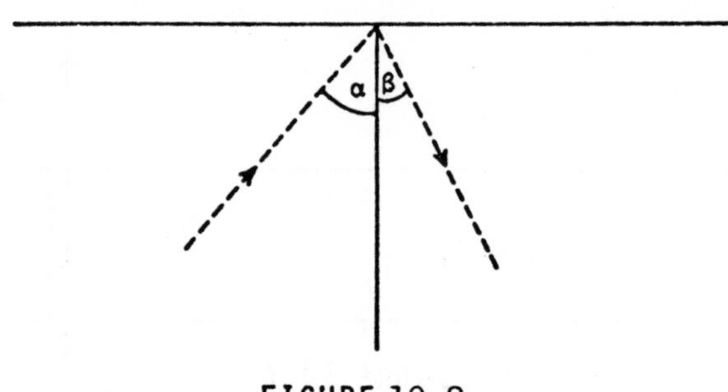

FIGURE 10.8

and where p and ξ are positive parameters associated with the nature of the repulsive force between two particles which have collided. The parameter p reflects the power of the repulsion, while the parameter ξ is a measure of how close the centers of two particles can come.

If and when a particle impacts on either the top or the bottom of the tube, or on the piston head, we will assume that it rebounds, as shown in Fig. 10.8, with

$$\beta = \alpha \pm \gamma. \tag{10.13}$$

The quantity γ is determined at random in the range $0 \le \gamma \le \pi/40$, subject to the restriction $0 \le \beta \le \pi/2$. If this last restriction is satisfied for both choices of sign in (10.13), then the sign is to be determined at random. If the incident speed is $|v_i|$, while the reflected speed is $|v_r|$, it will be assumed that

$$|v_r| = 0.2|v_i|,$$

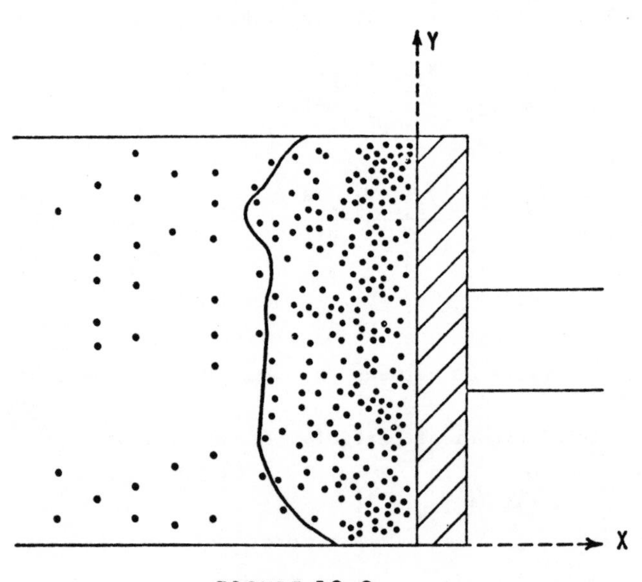

FIGURE 10.9

ich can be interpreted as a transference of kinetic energy from the molecules of e gas to the molecules of the container.

r the above simple formulation with $p = 1$ and $\xi = 0.1$, Fig. 10.9 shows the ock wave structure at time t_{160}, that is, after 1.6 seconds, when "greater an average" density is defined to mean that the distance between a particle and least five other particles is less than unity.

10.4 EXERCISES - CHAPTER 10

.1 Consider the two dimensional fluid example of Section 10.2 for each of $V = 50$ and $V = 1000$. Describe the resulting flows up to $t = 1.0$ when you use your computer to generate the velocity perturbations $\varepsilon_{i,1}$ and $\varepsilon_{i,2}$ for (10.2).

.2 Describe the qualitative changes in the flows of Exercise 10.1 for each of the following cases of parameter changes:

(a) $\alpha = 2$, $\beta = 4$

(b) $G = 1.5$, $H = 1.5$, $m_i = 0.01$

(c) $G = 1.1$, $H = 1.1$, $m_i = 3.0$

.3 Describe the appearance, decay, and reappearance of vortices for the flows of Exercises 10.1 and 10.2.

.4 Consider the two dimensional shock wave example of Section 10.3. Describe the resulting shock wave structure at t_{160} when you use your computer to generate the way particles enter the tube and to determine the quantities $\varepsilon_{i,1}$, $\varepsilon_{i,2}$ and $\pm\gamma$. How does your result compare with that shown in Fig. 10.9?

.5 Describe the qualitative changes in the resulting shock wave structure of Exercise 10.4 for each of the following cases of parameter changes:

(a) $p = 3$, $\xi = .12$

(b) $p = 5$, $\xi = .12$

(c) $p = 5$, $\xi = .01$

.6 Formulate a definition of liquid pressure. Why is the pressure greater at the bottom of a lake than at the top? Is the pressure at any point under water the same in all directions?

.7 Consider a square container full of liquid. Is there a greater concentration of liquid particles at the bottom of the container or at the top? Why? Would your answer be the same if "liquid" were replaced by "gas"? Why?

.8 When a heavy liquid (shown as the shaded area in the next figure) is set above a light liquid (shown as the unshaded area) in a square container, a reaction will occur in which the relative positions are interchanged. Simulate the reaction on a computer.

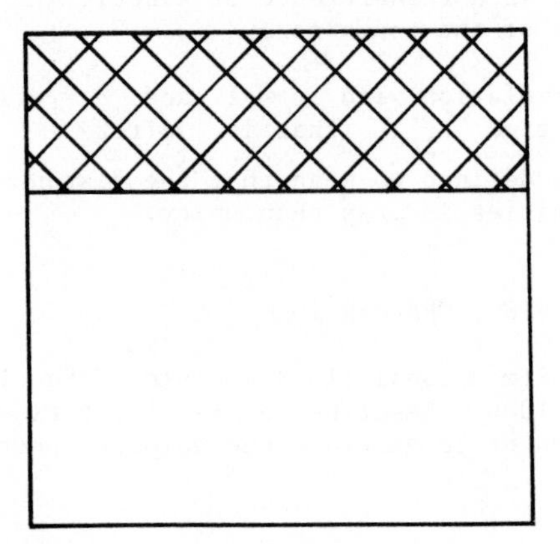

10.9 Simulate the development of a shock wave as a space module reenters the atmosphere from outer space. Why does the module develop a large amount o heat?

CHAPTER 11

Spinning Tops and Skaters

11.1 INTRODUCTION

There are many interesting and important phenomena in which the interactions of the smaller particles, which compose a larger body, are relatively insignificant when compared to the motion of the larger body itself. This is the case in all, so called, rigid bodies, typical of which are the motions of gyroscopes, spinning tops, and rotating planets. It is to such motions that we turn next. But, because of the vastness of the field, we will limit ourselves only to qualitative discussions of some of the more baffling types of behavior.

11.2 THE SPINNING TOP

Let us begin by developing more detailed notions related to angular momentum than those first introduced in Section 8.8.

Rotating objects have two very basic properties which can be discovered easily by experimenting with, for example, tops and gyroscopes. Let us state these by considering a rotating top. Assume that, as is usually the case, the top is physically symmetric about its center line, or axis, and that it rotates about this axis. Then, by virtue of its being in rotation, the top shows resistance to any force which might alter the direction of the axis of rotation. Indeed, this is what we actually mean when we say that the top has angular momentum by virtue of its being in rotation. In Section 8.8, when discussing the rotating bicycle wheel, we could just as easily have described its angular momentum in terms of resistance to change of the direction of the axle as resistance to change of the plane of motion. In the case of a top, however, the use of the axis for the discussion is more convenient because the top does not rotate in a single plane of motion.

In addition to the above property, observe that if one forces a new rotational motion on a spinning top, then the top will turn to align its axis with the axis of the imposed rotation. Indeed, this accommodation will allow the top to continue its spinning with as little resistance as possible.

A very interesting manifestation of the latter property is called the precession of a spinning top. It is an example in which the top rotates about its axis, while its axis rotates about a fixed vertical line. To explain this phenomenon, consider a spinning top as shown in Fig. 11.1. Initially, let $\overrightarrow{OP}$, the axis of the top, be

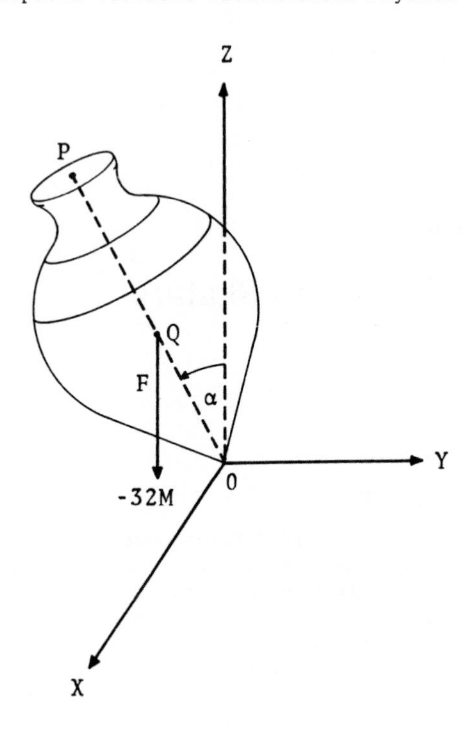

FIGURE 11.1

in the YZ plane and be inclined at a positive angle α with the Z-axis. Let Q be the centroid of the top and let M be its total mass. Then, in the units of Chapter 2, there is a force $\vec{F}$ whose magnitude is $-32M$ acting vertically at Q. Let us assume that the tip 0 of the top is stationary. If this is the case, the gravity is a force which makes Q, and hence axis $\overrightarrow{OP}$, rotate about 0. The axi of this rotational force is perpendicular to the plane determined by $\vec{F}$ and $\overrightarrow{OQ}$. The axis $\overrightarrow{OP}$ of the top then turns to try to align itself with this axis. But, just as soon as $\overrightarrow{OP}$ turns, $\overrightarrow{OQ}$ also turns, and so does the axis to the plane determined by $\vec{F}$ and $\overrightarrow{OQ}$. So, $\overrightarrow{OP}$ goes into a constant motion, called precessio trying to align itself with an axis which will not be caught.

11.3 ANGULAR VELOCITY

We want next to study the question of why a skater, who is spinning with his arms outstretched, can speed up his rotational motion merely by bringing his arms to hi sides. In order to do this it will be convenient to develop first the concept of angular velocity, that is, a type of velocity which has the same magnitude at ever point of the rotating skater. This is desirable since each particle of the skater's body turns through an angle of 2π radians in exactly the same time.

Let particle P be in rotation on the circle 0 of positive radius r, shown in Fig. 11.2. At time t_k, let the particle be at P, where θ_k, measured in radians, and s_k, the arc which subtends θ_k, are measured positively in the counterclockwise direction from the X-axis and negatively in the clockwise direction. Then, at t_k, the angular velocity w_k of P is defined by

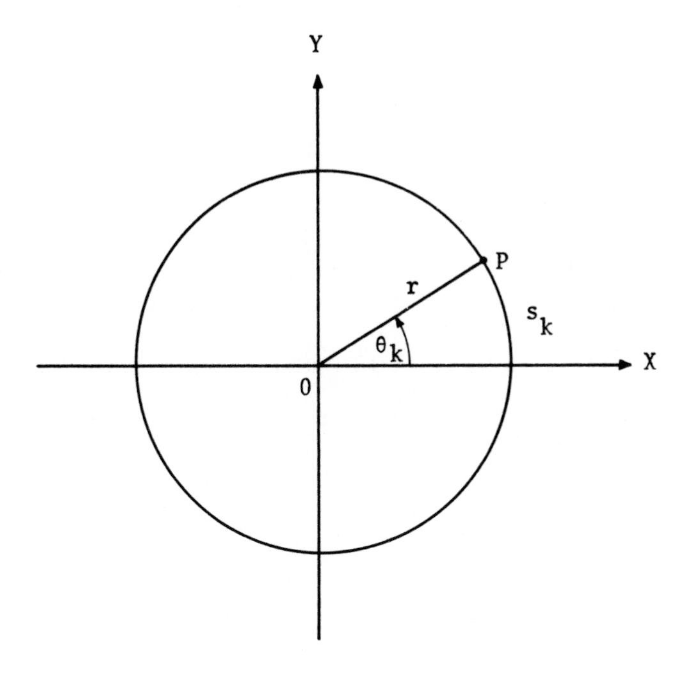

FIGURE 11.2

$$\frac{w_k+w_{k-1}}{2} = \frac{\theta_k-\theta_{k-1}}{\Delta t}, \quad k = 1,2,\ldots . \tag{11.1}$$

he reason why we do not require the use of vector notation in (11.1) is that, as n our study of the pendulum in Section 4.2, planar motion which is restricted to ie on a given circle is, essentially, motion in only one dimension, and in this ase only a plus or minus sign is needed to fix the direction.

ote that if the entire XY plane is in rigid rotation about 0, and if the nitial angular velocities of all the plane particles, except 0, are taken to ave the same value w_0, then at each t_k all the particles have the same angular elocity w_k. This follows directly from (11.1) and the observation that $\theta_k - \theta_{k-1}$ s the same for all particles, because we have assumed that the rotation is rigid.

ote also that it will be convenient to be able to relate a particle's angular elocity w_k with its linear velocity v_k, which can be done with the aid of the ollowing theorem.

heorem 11.1. Let particle P be in motion on a circle 0 of positive radius r. t t_k, let P's angular velocity w_k be given by (11.1) and its linear velocity $_k$ by

$$\frac{v_k+v_{k-1}}{2} = \frac{s_k-s_{k-1}}{\Delta t}. \tag{11.2}$$

Then, if

$$v_0 = rw_0, \tag{11.3}$$

it follows that

$$v_k = rw_k, \quad k = 1,2,\ldots . \tag{11.4}$$

<u>Proof</u>. Recall the basic relationship that

$$s = r\theta, \tag{11.5}$$

where, of course, θ must be given in radian measure. Then for $k = 1$, (11.1), (11.2) and (11.4) imply

$$\frac{w_1+w_0}{2} = \frac{\theta_1-\theta_0}{\Delta t} = \frac{1}{r}\,\frac{s_1-s_0}{\Delta t} = \frac{1}{r}\,\frac{v_1+v_0}{2}.$$

Hence,

$$w_1 r + w_0 r = v_1 + v_0.$$

By (11.3), then,

$$v_1 = rw_1.$$

Using the same steps as above,

$$v_n = rw_n$$

implies

$$v_{n+1} = rw_{n+1}.$$

Thus the theorem follows immediately by mathematical induction.

11.4 THE SPINNING SKATER

For the next phenomenon to be studied, we will require a definition of angular momentum for particles of a rotating body. In Section 8.8, this was defined as follows. At time t_k, if particle P_j of mass m_j is located at $\vec{r}_{j,k}$ and ha velocity $\vec{v}_{j,k}$, then its angular momentum is defined by (8.42), that is, by

$$\vec{L}_{j,k} = m_j(\vec{r}_{j,k}\times\vec{v}_{j,k}). \tag{11.6}$$

But, (11.5) was introduced to study the motion of particles which were all in the <u>same plane</u>. So, technically, we have to state now that (11.6) is defined also fo any particle in motion in three dimensions also, so that we can apply it to more general types of motions. But, it now becomes clear why vector methods were used to prove the law of conservation of angular momentum in Section 8.8, since that proof is also valid for three dimensional motion. Thus, the law of conservation angular momentum is still valid, i.e.,

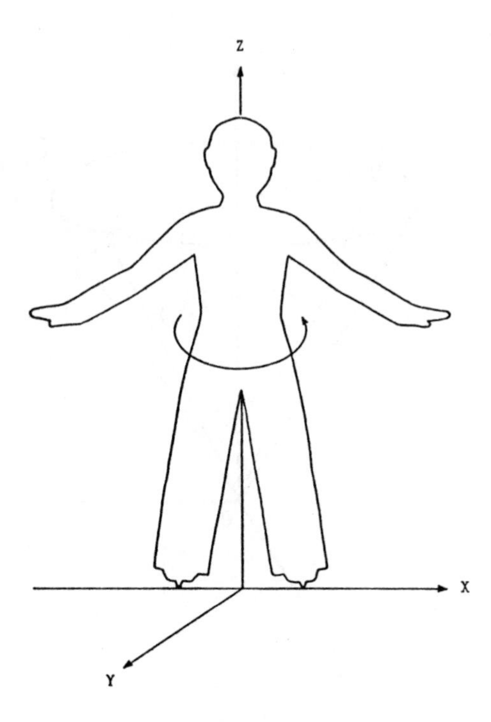

FIGURE 11.3

$$\vec{L}_j = \vec{L}_0, \quad j = 1,2,\ldots\ , \tag{11.7}$$

where $\vec{L}_j$ is defined as the angular momentum of the system by (8.43).

Consider now an ideal physical model of a skater, that is, one whose body is rigid and is symmetrical about an axis and whose body particles all have the same mass m. As shown in Fig. 11.3, the axis of symmetry is taken to be the Z-axis. Assume that the skater rotates about his axis, so that the particles not on the axis rotate in circles whose planes are all perpendicular to the Z-axis. Assume that each such particle has the same initial angular velocity w_0, so that each particle has the same angular velocity w_k at time t_k, by virtue of the assumption of rigidity. For simplicity, we will assume also that the skater's angular velocity is constant, and equal to w, for all t_k.

Consider now a pair of body particles P_i and P_j which are symmetric about the Z-axis, as shown in Fig. 11.4, and let us determine the angular momentum contribution of this pair. At t_k, let P_i be at $\vec{r}_{i,k}$ with velocity $\vec{v}_{i,k}$ while P_j is at $\vec{r}_{j,k}$ with velocity $\vec{v}_{j,k}$. Then, from (11.6)

$$\vec{L}_{i,k} = m(\vec{r}_{i,k} \times \vec{v}_{i,k}) \tag{11.8}$$

$$\vec{L}_{j,k} = m(\vec{r}_{j,k} \times \vec{v}_{j,k}). \tag{11.9}$$

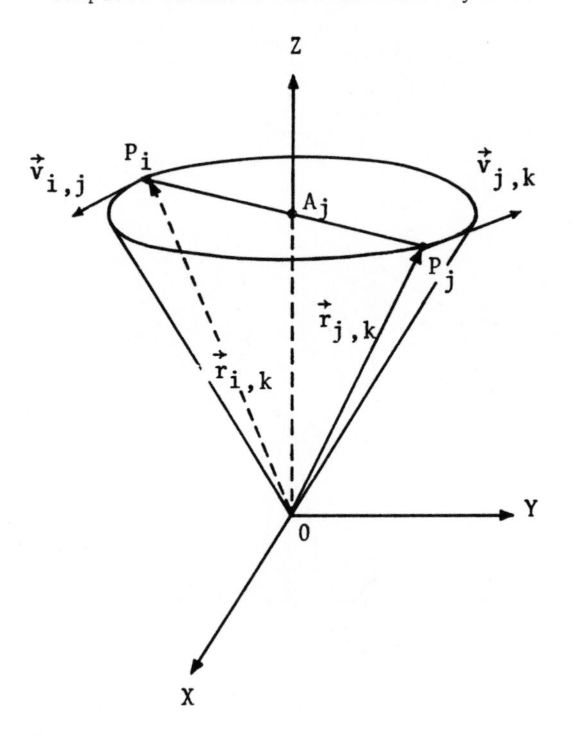

FIGURE 11.4

However, if A_j is the center of the circle of rotation for both P_i and P_j, then

$$\vec{r}_{i,k} = \overrightarrow{OA_j} + \overrightarrow{A_jP_i} \tag{11.10}$$

$$\vec{r}_{j,k} = \overrightarrow{OA_j} + \overrightarrow{A_jP_j} \tag{11.11}$$

But, (11.8) and (11.10) imply

$$\vec{L}_{i,k} = m(\overrightarrow{OA_j}+\overrightarrow{A_jP_i}) \times \vec{v}_{i,k} = m\overrightarrow{OA_j} \times \vec{v}_{i,k} + m\overrightarrow{A_jP_i} \times \vec{v}_{i,k},$$

while (11.9) and (11.11) yield

$$\vec{L}_{j,k} = m\overrightarrow{OA_j} \times \vec{v}_{j,k} + m\overrightarrow{A_jP_j} \times \vec{v}_{j,k}.$$

Hence, since

$$\vec{v}_{i,k} + \vec{v}_{j,k} = \vec{0},$$

one has

$$\vec{L}_{i,k} + \vec{L}_{j,k} = m(\overrightarrow{A_jP_j}-\overrightarrow{A_jP_i}) \times \vec{v}_{j,k}.$$

Thus, if the skater's body has a total of 2N symmetric particles in motion, then the total angular momentum at t_k is

$$\vec{L}_k = m \sum_{j=1}^{N} (\overrightarrow{A_jP_j} - \overrightarrow{A_jP_i}) \times v_{j,k} \tag{11.12}$$

r

$$\vec{L}_k = m \sum_{j=1}^{N} (\overrightarrow{P_iP_j} \times \vec{v}_{j,k}). \tag{11.13}$$

ote that even if there are particles on the axis of rotation, (11.13) still gives he correct total angular momentum for the skater, since the angular momentum of ach particle on the axis is $\vec{0}$, by (11.6).

inally, let us examine the magnitude of $\vec{L}_k$. By the conservation of angular omentum, $\vec{L}_k$ is the same vector at each value of k. Hence, the magnitude of $_k$ does not change with k and is given by

$$|\vec{L}_k| = m\left| \sum_{j=1}^{N} (\overrightarrow{P_iP_j} \times \vec{v}_{j,k}) \right|. \tag{11.14}$$

ut, each of the vectors $\overrightarrow{P_iP_j} \times \vec{v}_{j,k}$ has the same direction, as is evident from ig. 11.4 and the definition of the cross product, so that

$$|\vec{L}_k| = m \sum_{j=1}^{N} |\overrightarrow{P_iP_j} \times \vec{v}_{j,k}|. \tag{11.15}$$

hus,

$$|\vec{L}_k| = m \sum_{j=1}^{N} |\overrightarrow{P_iP_j}| \cdot |\vec{v}_{j,k}| \sin\frac{\pi}{2} = 2m \sum_{j=1}^{N} |\overrightarrow{A_jP_j}| \cdot |\overrightarrow{A_jP_j}| \cdot w,$$

o that

$$|\vec{L}_k| = 2mw \sum_{j=1}^{N} |\overrightarrow{A_jP_j}|^2. \tag{11.16}$$

e are now ready to explain the rotating skater's dramatic increase in velocity hen he brings his outstretched arms to his sides, for when he does this, the quan- ities $|\overrightarrow{A_jP_j}|$ corresponding to the particles in his arms all decrease. However, $|\vec{L}_k|$ in (11.16) is invariant, that is, it does not change. So, to compensate, hen the summation on the right-hand side decreases, w must increase, and the henomenon is now partially explained. We say partially, because the increase in seems to be exceptionally great. But this too follows from (11.16), because w ust compensate for the squares of the values $|\overrightarrow{A_jP_j}|$. For example, if $|\overrightarrow{A_jP_j}|$ hanges from 3 to 1, then $|\overrightarrow{A_jP_j}|^2$ changes from 9 to 1. Thus, for example, f $|\overrightarrow{A_jP_j}|$ changes from 3 to 1, w must increase in the ratio 1 to 9, and ow the entire phenomenon is understood.

11.5 EXERCISES - CHAPTER 11

11.1 Try to balance a gyroscope on a string when the gyroscope is not spinning. Explain your failure.

11.2 Balance a spinning gyroscope on a string in such a fashion that the gyroscope's axis is not vertical. Does the gyroscope precess? Why? Next, sn the string in order to eject the gyroscope into the air. Does the gyrosco precess while it is in the air? Why?

11.3 A fencing sword, which is symmetric and balanced, is to be thrown into the air and caught when it comes down. What would be the advantage of throwin it with as much spin as possible?

11.4 A particle P is in rotation on a circle whose radius r is 10 cm long and whose center is the origin. Initially, the particle has $w_0 = 0$. For $\Delta t = 1$ sec, and for each of the following laws of motion, find the particle's angular velocity at t_1, t_2, t_3, t_4, t_5.

(a) $s_k = 2$

(b) $s_k = 5t_k^2$

(c) $s_k = 10 - 5t_k^2$

(d) $s_k = t_k^3$.

11.5 Prove that

$$w_1 = \frac{2}{\Delta t}[\theta_1 - \theta_0] - w_0$$

$$w_k = \frac{2}{\Delta t}[\theta_k + (-1)^k\theta_0 + 2\sum_{j=1}^{k-1}(-1)^j\theta_{k-j}] + (-1)^k w_0; \quad k \geq 2.$$

11.6 Let particle P be in motion as shown in Fig. 11.2. Define P's angular acceleration α_k by

$$\alpha_k = \frac{w_{k+1} - w_k}{\Delta t}, \quad k = 0,1,\ldots .$$

Calculate $\alpha_0, \alpha_1, \alpha_2, \alpha_3, \alpha_4$ for each motion of Exercise 11.4.

11.7 From the definition of angular acceleration of Exercise 11.6, prove that:

$$\alpha_0 = \frac{2}{(\Delta t)^2}[\theta_1 - \theta_0 - w_0\Delta t]$$

$$\alpha_1 = \frac{2}{(\Delta t)^2}[\theta_2 - 3\theta_1 + 2\theta_0 + w_0\Delta t]$$

$$\alpha_{k-1} = \frac{2}{(\Delta t)^2}\{\theta_k - 3\theta_{k-1} + 2(-1)^k\theta_0 + 4\sum_{j=2}^{k-1}[(-1)^j\theta_{k-j}] + (-1)^k w_0\Delta t\}, \ k \geq$$

1.8 For each motion of Exercise 11.4, assume that $v_0 = 0$ and find v_1, v_2, v_3, v_4, v_5.

1.9 Model an idealized gyroscope and estimate the magnitude of its angular momentum vector if it has a total mass of 100 grams and an angular velocity of 2π radians per second.

1.10 Model an idealized top and estimate the magnitude of its angular momentum vector if it has a total mass of 100 grams and an angular velocity of 2π radians per second.

CHAPTER 12

The Galilean Principle of Relativity

12.1 INTRODUCTION

At first, it might seem reasonable to expect that a person experimenting in a mov ing train might get results which are fundamentally different than those of a per son who is not in motion at all. And this can, indeed, be the case, for when a train is accelerating or decelerating, for example, even a ball rolled across the aisle on the train floor will not roll in a straight line. But, when the train moves with a constant velocity, the same ball will roll in a straight line. Indee when on a train which is moving at a constant velocity, we know that if we pull down all the shades it becomes quite difficult to deduce that we are, in fact, in motion. Perhaps the noise of the rotating wheels or an occasional rocking of the car will signal this motion, but when these are not apparent, there seems to be n clue at all. Physically, this absence of clues suggests that all the laws of phy ics are the same both at rest and when one is in motion at a constant velocity. This beautiful result is called the Galilean principle of relativity, or symmetry We turn next, then, to its precise formulation and to a proof of its validity.

12.2 THE GALILEAN PRINCIPLE

In a fixed XYZ coordinate system, let a force acting on particle P of mass m be given by $\vec{F}_k$ at time t_k. If the resulting acceleration of P is given by $\vec{a}_k$ then we have assumed that $\vec{F}_k$ and $\vec{a}_k$ are related by

$$\vec{F}_k = m\vec{a}_k. \tag{12.1}$$

Consider now a second coordinate system X'Y'Z' which moves with a constant velo ity relative to the XYZ system. In the X'Y'Z' system, let the force on P b denoted by $\vec{F}'_k$, while the acceleration of P is denoted by $\vec{a}'_k$. Then the Galile principle of relativity states that in the X'Y'Z' system the force and the acce eration are related by

$$\vec{F}'_k = m\vec{a}'_k. \tag{12.2}$$

Thus, the relationships between force and acceleration have exactly the same form in both systems, which is the meaning of the commonly used statement that the law physics are the same in both coordinate systems.

order to prove the above result, it is sufficient to show that

$$F_{k,x} = ma_{k,x}, \quad F_{k,y} = ma_{k,y}, \quad F_{k,z} = ma_{k,z} \tag{12.3}$$

nply

$$F_{k,x'} = ma_{k,x'}, \quad F_{k,y'} = ma_{k,y'}, \quad F_{k,z'} = ma_{k,z'}, \tag{12.4}$$

nce (12.1) and (12.2) are vector equations. For simplicity, we will proceed to ow only that

$$F_{k,x} = ma_{k,x} \tag{12.5}$$

nplies

$$F_{k,x'} = ma_{k,x'}, \tag{12.6}$$

nce the other implications of (12.4) follow from (12.3) in exactly the same way.

ssume then that an X'-axis is in uniform motion relative to an X-axis, so that the ordinates of the two systems are related by

$$x' = x - ct_k, \quad k = 0,1,\dots , \tag{12.7}$$

ere c is a nonzero constant. Equation (12.7) is called a Galilean transforma-on, from which we derive the name of the relativistic principle being developed. w, if $v_{0,x}$ is the initial velocity of P's motion in the X system, let its itial velocity $v_{0,x'}$ in the X' system be given by

$$v_{0,x'} = v_{0,x} - c. \tag{12.8}$$

en, from (12.7), (12.8) and Theorem 2.2,

$$\begin{aligned} v_{1,x} &= \frac{2}{\Delta t}[(x_1'+ct_1) - (x_0'+ct_0)] - v_{0,x} = \frac{2}{\Delta t}[x_1' - x_0'] - v_{0,x} + 2c \\ &= \frac{2}{\Delta t}[x_1' - x_0'] - (v_{0,x'}) + c = v_{1,x'} + c, \end{aligned} \tag{12.9}$$

ile, for $k \geq 2$,

$$\begin{aligned} v_{k,x} &= \frac{2}{\Delta t}\Big\{(x_k'+ct_k) + (-1)^k(x_0'+ct_0) \\ &\qquad + 2\sum_{j=1}^{k-1}[(-1)^j(x_{k-j}'+ct_{k-j})]\Big\} + (-1)^k v_{0,x} \\ &= \frac{2}{\Delta t}\Big\{x_k' + (-1)^k x_0' + 2\sum_{j=1}^{k-1}[(-1)^j x_{k-j}']\Big\} + (-1)^k v_{0,x} \\ &\qquad + \frac{2c}{\Delta t}\Big\{t_k + (-1)^k t_0 + 2\sum_{j=1}^{k-1}[(-1)^j t_{k-j}]\Big\}. \end{aligned} \tag{12.10}$$

But,

$$t_k + (-1)^k t_0 + 2\sum_{j=1}^{k-1}[(-1)^j t_{k-j}] = \begin{cases} 0, & k \text{ even} \\ \Delta t, & k \text{ odd.} \end{cases} \qquad (12.1$$

Thus, for k even,

$$v_{k,x} = \frac{2}{\Delta t}\left\{x'_k + x'_0 + 2\sum_{j=1}^{k-1}[(-1)^j x'_{k-j}]\right\} + (v_{0,x'}+c) = v_{k,x'} + c, \qquad (12.1$$

while, for k odd,

$$v_{k,x} = \frac{2}{\Delta t}\left\{x'_k - x'_0 + 2\sum_{j=1}^{k-1}[(-1)^j x'_{k-j}]\right\} - v_{0,x} + 2c = v_{k,x'} + c. \qquad (12.1$$

Thus, one has from (3.6), (12.9), (12.12) and (12.13) that

$$ma_{k,x} = m\frac{v_{k+1,x}-v_{k,x}}{\Delta t} = m\frac{(v_{k+1,x'}+c)-(v_{k,x'}+c)}{\Delta t} = ma_{k,x'}. \qquad (12.1$$

Finally, assuming that the force applied in both systems is the same, that is, $F_x = F_{x'}$, then

$$F_x = ma_{k,x}$$

implies, by (12.14), that

$$F_{x'} = ma_{k,x'},$$

and the principle is established.

12.3 REMARKS

Usually, the name of Einstein is thought to be inseparable from the concept of relativity. Let us then remark now that we have, in fact, proved a less comprehensive principle than Einstein's, and for two basic reasons: (1) it is by far t easier of the two to establish and, (2) it is sufficient to cover all previously discussed phenomena. It is only when one studies, for example, atomic particles which move at very high speeds, say, close to the speed of light, that one finds obvious disagreement between experimental results and Newtonian theory. At that time, the Einstein theory becomes essential.

12.4 EXERCISES - CHAPTER 12

12.1 Why will a ball rolled across the aisle on a train floor not roll in a straight line when the train is accelerating or decelerating?

12.2 If an elevator is accelerating vertically at -32 ft/sec^2 and a man in th elevator drops a ball from a position of rest, what will he observe to be the motion of the ball? If the elevator is rising vertically at a constar velocity, what will he observe?

2.3 Let an X'-axis be in uniform motion relative to an X-axis. Let the coordinates of the axes be related by (12.7). For $\Delta t = 1$ and $k = 0,1,2,3$, determine the x' coordinate which corresponds to x for each of the following cases.

(a) $x = 0, \quad c = 1$

(b) $x = 0, \quad c = -1$

(c) $x = 2, \quad c = -3$

(d) $x = -5, \quad c = 6.$

2.4 Let an X'-axis be in uniform motion relative to an X-axis. Let the coordinates be related by (12.7). For $\Delta t = 1$ and $k = 0,1,2,3$, determine the x coordinates which correspond to x' for each of the following cases.

(a) $x' = 0, \quad c = 1$

(b) $x' = 0, \quad c = -1$

(c) $x' = 1, \quad c = -2$

(d) $x' = -3, \quad c = 4.$

2.5 Let an X'-axis be in uniform motion relative to an X-axis. Let the initial velocity of a particle P in the X system be $v_{0,x}$. Determine its initial velocity $v_{0,x'}$ in the X' system by (12.8) for each of the following cases.

(a) $v_{0,x} = 0, \quad c = 1$

(b) $v_{0,x} = 0, \quad c = -1$

(c) $v_{0,x} = 5, \quad c = 2$

(d) $v_{0,x} = -3, \quad c = -4.$

2.6 Let an X'-axis be in uniform motion relative to an X-axis. Let the initial velocity of a particle P in the X' system be $v_{0,x'}$. Determine its initial velocity $v_{0,x}$ in the X system by (12.8) for each of the following cases.

(a) $v_{0,x'} = 0, \quad c = 1$

(b) $v_{0,x'} = 0, \quad c = -1$

(c) $v_{0,x'} = 3, \quad c = 2$

(c) $v_{0,x'} = -3, \quad c = -7.$

12.7 Show that (12.11) is true for each $k = 0,1,2,3,4,5,6,7$. Then prove (12.11) is correct for all values of k.

APPENDIX A

Fortran Program for the Harmonic Oscillator Example of Section 4.4

```
      PARAMETER N=75
      IMPLICIT DOUBLE PRECISION(A-H,M,O-Z)
      DIMENSION X(201),V(201),KNUM(201)
      DATA PI/3.141592653599793D+00/
 2001 FORMAT(1H1,4X,'K',8X,'X',10X,'V',3(9X,'K',8X,'X',10X,'V'))
 2002 FORMAT(4(3X,I5,2F11.7))
C  COMPUTE PROGRAM CONSTANTS
      H=1.D-01
      H2=H*H
      D=4.D+00+H2
      A=(4.D+00-H2)/D
      B=4.D+00*H/D
      X(201)=PI/4.D+00
      V(201)=0.D+00
C  COMPUTE ITERATES, 200 AT A TIME
      DO 300 KPASS=1,N
      X(1)=X(201)
      V(1)=V(201)
      DO 100 J=2,201
      JM1=J-1
      X(J)=A*X(JM1)+B*V(JM1)
      V(J)=-B*X(JM1)+A*V(JM1)
  100 CONTINUE
C  OUTPUT CURRENT GROUP OF ITERATES
      PRINT 2001
      INDEX=(KPASS-1)*200
      DO 150 J=2,201
      KNUM(J)=INDEX+J-1
  150 CONTINUE
      DO 200 JL=2,51
      JU=JL+150
      PRINT 2002,(KNUM(J),X(J),V(J),J=JL,JU,50)
  200 CONTINUE
C  COMPUTE NEW SET OF ITERATES
  300 CONTINUE
      STOP
      END
```

APPENDIX B

Fortran Program for the Wave Interaction Example of Section 5.3

```
      PARAMETER N=101,NM1=N-1,JV=7,JV2=97
      PARAMETER KMAX=3000,IPRINT=50
      REAL MASS
      DIMENSION X(N),Y(N),VO(N),V(N,2),A(N,2),R(N),T(N),JNUM(N)
      DATA MASS,TO,G/0.01,10.0,0.0/
      DATA DX,EPS/0.02,0.02/
      DATA DT/0.0001/
C
 1001 FORMAT(10F5.0)
 2000 FORMAT(1H1,' PROGRAM CONSTANTS'/5X,'MASS= ',F5.2,5X,'TO= ',F5.1,
     15X,'G= ',F4.1/5X,'DX= ',F5.2,5X,'DT= ',F7.5,5X,'EPS= ',F5.2)
 2001 FORMAT(1H1,' TIMESTEP= ',I4/8X,'K',11X,'Y(K)',3(9X,'K',10X,'Y(K)'))
     1)
 2002 FORMAT(4(5X,I5,5X,F10.5))
 2003 FORMAT(/' INITIAL VELOCITIES--LARGE WAVE'/5X,6F10.1)
 2004 FORMAT(/' INITIAL VELOCITIES--SMALL WAVE'/5X,6F10.1)
 2005 FORMAT(/' ALPHA= ',F6.3)
C
      DXSQ=DX**2
      DT2=DT/2.0
      READ 1001,(VO(J),J=2,JV)
      READ 1001,(VO(J),J=JV2,NM1)
    1 READ 1001,ALPHA
      IF(ALPHA.LT.0)STOP
C  INITIALIZE POSITIONS, VELOCITIES,ACCELERATIONS
      K=0
      DO 10 J=1,N
      X(J)=(J-1)*DX
      Y(J)=0.0
      V(J,2)=VO(J)
      A(J,2)=0.0
      JNUM(J)=J-1
   10 CONTINUE
C  PRINT INITIAL CONDITIONS, PARAMETERS
      PRINT 2000,MASS,TO,G,DX,DT,EPS
      PRINT 2003,(VO(J),J=2,JV)
      PRINT 2004,(VO(J),J=JV2,NM1)
      PRINT 2005,ALPHA
```

```
      GO TO 85
C  UPDATE VELOCITIES, ACCELERATIONS
   39 K=K+1
      DO 40 J=1,N
      V(J,1)=V(J,2)
      A(J,1)=A(J,2)
   40 CONTINUE
C  COMPUTE TENSION BETWEEN NEIGHBORING PARTICLES
      DO 50 J=1,NM1
      JP1=J+1
      R(J)=SQRT(DXSQ+(Y(JP1)-Y(J))**2)
   50 CONTINUE
      OME=1.0-EPS
      DO 60 J=1,NM1
      RDX=R(J)/DX
      T(J)=T0*(OME*RDX+EPS*RDX**2)
   60 CONTINUE
C  COMPUTE ACCELERATIONS
      DO 70 J=2,NM1
      JP1=J+1
      JM1=J-1
      A(J,2)=T(J)*(Y(JP1)-Y(J))/R(J)-T(JM1)*(Y(J)-Y(JM1))/R(JM1)-ALPHA*
     1V(J,2)-MASS*G
      A(J,2)=A(J,2)/MASS
   70 CONTINUE
C  COMPUTE VELOCITIES, POSITIONS--CURRENT TIMESTEP
      DO 80 J=2,NM1
      V(J,2)=V(J,1)+DT*A(J,2)
      Y(J)=Y(J)+DT2*(V(J,2)+V(J,1))
   80 CONTINUE
C
      IF(MOD(K,IPRINT).NE.0)GO TO 100
C  PRINT
   85 PRINT 2001,K
      DO 90 JL=2,26
      JU=JL+75
      PRINT 2002,(JNUM(J),Y(J),J=JL,JU,25)
   90 CONTINUE
  100 IF(K.LT.KMAX)GO TO 39
      GO TO 1
      END
```

APPENDIX C

Fortran Program for the Orbit Calculation of Section 7.7

```
      DATA X3,Y3,VX3,VY3/0.5,0.0,0.0,1.63/
      DATA MAXIT,EPS,DT/50,1.E-04,1.E-03/
 2001 FORMAT(1H1,' ONE-BODY ORBIT'//5X,'N',10X,'X',14X,'Y',13X,'VX',
     113X,'VY',14X,'E'//)
 2002 FORMAT(2X,I6,5(5X,F10.4))
 2003 FORMAT('0CONVERGENCE FAILURE IN ',I2,' ITERATIONS AT TIMESTEP ',
     1I6)

      PRINT 2001
      IPRINT=10
      NMAX=350000
      DT2=DT/2.
      R2=0.5
      N=0
      GO TO 30
C  UPDATE POSITIONS, VELOCITIES--PREVIOUS TIMESTEP
   10 N=N+1
      IF(N.GT.NMAX)STOP
      X1=X3
      Y1=Y3
      VX1=VX3
      VY1=VY3
      R1=R2
      J=0
   15 J=J+1
      IF(J.GT.MAXIT)GO TO 35
C  UPDATE PREVIOUS ITERATES, COMPUTE CURRENT ITERATES
      X2=X3
      Y2=Y3
      VX2=VX3
      VY2=VY3
      X3=X1+DT2*(VX2+VX1)
      Y3=Y1+DT2*(VY2+VY1)
      R2=SQRT(X3**2+Y3**2)
      RR=R1*R2*(R1+R2)
      VX3=VX1-DT*(X3+X1)/RR
      VY3=VY1-DT*(Y3+Y1)/RR
```

```
C  TEST FOR CONVERGENCE
      IF(ABS(X3-X2).GT.EPS)GO TO 15
      IF(ABS(Y3-Y2).GT.EPS)GO TO 15
      IF(ABS(VX3-VX2).GT.EPS)GO TO 15
      IF(ABS(VY3-VY2).GT.EPS)GO TO 15
C  CONVERGENCE
   30 IF(N.NE.0.AND.MOD(N,IPRINT).NE.0)GO TO 10
      E=0.5*(VX3**2+VY3**2)-1./R2
      PRINT 2002,N,X3,Y3,VX3,VY3,E
      GO TO 10
C  CONVERGENCE FAILURE
   35 PRINT 2003,MAXIT,N
      STOP
      END
```

APPENDIX D

Fortran Program for Three-Body Problem of Section 8.2

```
      IMPLICIT DOUBLE PRECISION (A-H,O-Z)
      DOUBLE PRECISION MASS(3)
      DIMENSION XO(3),YO(3),VXO(3),VYO(3),X(3,3),Y(3,3),VX(3,3),VY(3,3),
     * FX(3),FY(3),R(3,2),GM(3)
      DATA MAXIT/150/,EPS/1.D-10/,G/1.D-02/
C
C THE ABOVE DATA IS DEFINED AS FOLLOWS
C     MAXIT = MAX. NO. OF NEWTON ITERATIONS TO BE ALLOWED IN COM-
C             PUTING POSITIONS AND VELOCITIES AT EACH TIME STEP
C       EPS = CONVERGENCE TOLERANCE IN NEWTON ITERATION
C         G = GRAVITATIONAL CONSTANT
C
   99 FORMAT(16D5.0)
   98 FORMAT(2I5,14D5.0)
   97 FORMAT('1DT =',D11.4,5X,'OMEGA =',D11.4,3X,'VX,VY =',6D10.3/
     * 1X,'MASS ='3D11.4/)
   96 FORMAT('0NEWTON ITERATION FAILED AFTER',I5,2X,'ITERATIONS')
   95 FORMAT('0NEWTON ITERATION FOR TIME STEP',I5)
   94 FORMAT(' R,FX,FY',2X,9D13.6)
   93 FORMAT(1X,6D15.7)
C
C READ PROBLEM-DEFINING PHYSICAL DATA
C       XO(I),YO(I),I=1,2,3 = INITIAL PARTICLE POSITIONS
C     VXO(I),VYO(I),I=1,2,3 = INITIAL PARTICLE VELOCITIES
C            MASS(I),I=1,2,3 = PARTICLE MASSES
C
    3 READ(5,99,END=40) (XO(I),YO(I),I=1,3),(VXO(I),VYO(I),I=1,3),
     * (MASS(I),I=1,3)
C
C READ ADDITIONAL COMPUTATIONAL DATA
C      NMAX = MAXIMUM NO. OF TIME STEPS FOR THIS DATA CASE
C     INCPR = TIME STEP INCREMENT FOR PRINTING OF RESULTS
C     OMEGA = OVERRELAXATION FACTOR IN NEWTON'S METHOD
C        DT = TIME STEP SIZE
C       END = INPUT CONTROL VARIABLE
C             END.EQ.0 IMPLIES MORE CARDS OF THIS TYPE WILL FOLLOW
C                  FOR THE SAME PHYSICAL DATA INPUT ABOVE
C             END.NE.0 IMPLIES NOT
```

```
    5 READ(5,98)NMAX,INCPR,OMEGA,DT,END
      WRITE(6,97)DT,OMEGA,(VXO(I),VYO(I),I=1,3),(MASS(I),I=1,3)

C  COMPUTE STATIC DATA-DEPENDENT PROGRAM VARIABLES AND INITIALIZE
C     DYNAMIC POSITION AND VELOCITY VECTORS

C  DEFINITIONS OF X,Y,VX,VY ARRAYS

C   FOR I=1,2,3
C      X(I,1) = X-COMPONENT OF POSITION OF PARTICLE I AT PREVIOUS
C               TIME STEP
C      X(I,2) = SAME AS ABOVE, EXCEPT AT CURRENT TIME STEP AND
C               PREVIOUS NEWTON ITERATION
C      X(I,3) = SAME AS ABOVE, EXCEPT AT CURRENT NEWTON ITERATION
C     VX(I,1) = X-COMPONENT OF VELOCITY OF PARTICLE I
C     VX(I,2) =   WITH DEFINITION OF SECOND SUBSCRIPT
C     VX(I,3) =     SIMILAR TO THAT GIVEN FOR X ABOVE
C      Y(I,1) = SAME AS ABOVE
C      Y(I,2) =   EXCEPT FOR
C      Y(I,3) =     Y-COMPONENTS
C     VY(I,1) =       OF POSITION
C     VY(I,2) =         AND VELOCITY
C     VY(I,3) =           OF PARTICLE I

      OME1=1.-OMEGA
      DT2=.5*DT
      DO 8 I=1,3
      GM(I)=G*MASS(I)
      X(I,3)=XO(I)
      Y(I,3)=YO(I)
      VX(I,3)=VXO(I)
    8 VY(I,3)=VYO(I)

C  COMPUTE INITIAL DISTANCES BETWEEN PARTICLES

      CALL RR(R(1,2))

C  PRINT OUT INITIAL PARTICLE POSITIONS

      CALL PRINT(0,X,Y)

C  BEGIN TIME STEP LOOP

      N=0
   10 N=N+1

C  UPDATE DISTANCES BETWEEN PARTICLES,PARTICLE POSITIONS, AND
C     PARTICLE VELOCITIES FOR PREVIOUS TIMESTEP

      DO 12 I=1,3
      R(I,1)=R(I,2)
      X(I,1)=X(I,3)
      Y(I,1)=Y(I,3)
      VX(I,1)=VX(I,3)
   12 VY(I,1)=VY(I,3)

C  BEGIN NEWTON ITERATION LOOP

      DO 25 J=1,MAXIT
```

```
C  UPDATE PREVIOUS ITERATES FOR POSITIONS AND VELOCITIES AND COMPUTE
C     CURRENT ITERATES FOR POSITIONS

      DO 14 I=1,3
      X(I,2)=X(I,3)
      VX(I,2)=VX(I,3)
      X(I,3)=OME1*X(I,2)+OMEGA*(DT2*(VX(I,2)+VX(I,1))+X(I,1))
      Y(I,2)=Y(I,3)
      VY(I,2)=VY(I,3)
   14 Y(I,3)=OME1*Y(I,2)+OMEGA*(DT2*(VY(I,2)+VY(I,1))+Y(I,1))

C  COMPUTE DISTANCES AND FORCES BETWEEN PARTICLES FOR CURRENT VALUES
C    OF POSITION ITERATES

      CALL RR(R(1,2))
      CALL FXY

C  COMPUTE CURRENT ITERATES FOR VELOCITIES

      DO 16 I=1,3
      VX(I,3)=OME1*VX(I,2)+OMEGA*(DT*FX(I)+VX(I,1))
   16 VY(I,3)=OME1*VY(I,2)+OMEGA*(DT*FY(I)+VY(I,1))

C  TEST FOR CONVERGENCE OF NEWTON ITERATION
      DO 18 I=1,3
      IF(ABS(X(I,3)-X(I,2)).GT.EPS)GO TO 25
      IF(ABS(Y(I,3)-Y(I,2)).GT.EPS)GO TO 25
      IF(ABS(VX(I,3)-VX(I,2)).GT.EPS)GO TO 25
   18 IF(ABS(VY(I,3)-VY(I,2)).GT.EPS)GO TO 25
      GO TO 30
   25 CONTINUE

C  END OF NEWTON ITERATION LOOP
C     WRITE CONVERGENCE FAILURE MESSAGE AND GO TO NEXT DATA CASE

      WRITE (6,96)MAXIT
      GO TO 35

C  TEST FOR PRINTING OF POSITIONS AT CURRENT TIME STEP

   30 IF(MOD(N,INCPR).EQ.0)CALL PRINT(N,X,Y)

C  TEST FOR END OF TIME STEP LOOP FOR CURRENT DATA CASE

      IF(N.LT.NMAX)GO TO 10

C  END OF TIME STEP LOOP.  TEST FOR LAST COMPUTATIONAL DATA CASE
C    FOR CURRENT PHYSICAL DATA.

   35 IF(END.GT.0)GO TO 3
      GO TO 5

C  TERMINATION POINT FOR PROGRAM.  CONTROL REACHES HERE UPON
C     ATTEMPTING TO READ PAST LAST DATA CARD.

   40 STOP
```

```
C  INTERNAL SUBROUTINE FOR COMPUTING DISTANCES BETWEEN PARTICLES

      SUBROUTINE RR(R)
      DIMENSION R(3)
      R(1)=SQRT((X(1,3)-X(2,3))**2+(Y(1,3)-Y(2,3))**2)
      R(2)=SQRT((X(1,3)-X(3,3))**2+(Y(1,3)-Y(3,3))**2)
      R(3)=SQRT((X(2,3)-X(3,3))**2+(Y(2,3)-Y(3,3))**2)
      RETURN

C  INTERNAL SUBROUTINE FOR COMPUTING FX(I),FY(I),I=1,2,3
C     WHERE FX(I) = X-COMPONENT OF TOTAL FORCES ACTING ON PARTICLE I
C                   DIVIDED BY MASS OF PARTICLE I
C           FY(I) = SAME AS ABOVE WITH Y-COMPONENT

      SUBROUTINE FXY
      DIMENSION D(3)
      DO 2 I=1,3
    2 D(I)=R(I,1)*R(I,2)*(R(I,1)+R(I,2))
      TX1=(X(1,3)+X(1,1)-X(2,3)-X(2,1))/D(1)
      TY1=(Y(1,3)+Y(1,1)-Y(2,3)-Y(2,1))/D(1)
      TX2=(X(1,3)+X(1,1)-X(3,3)-X(3,1))/D(2)
      TY2=(Y(1,3)+Y(1,1)-Y(3,3)-Y(3,1))/D(2)
      TX3=(X(2,3)+X(2,1)-X(3,3)-X(3,1))/D(3)
      TY3=(Y(2,3)+Y(2,1)-Y(3,3)-Y(3,1))/D(3)
      FX(1)=-GM(2)*TX1-GM(3)*TX2
      FY(1)=-GM(2)*TY1-GM(3)*TY2
      FX(2)= GM(1)*TX1-GM(3)*TX3
      FY(2)= GM(1)*TY1-GM(3)*TY3
      FX(3)= GM(1)*TX2+GM(2)*TX3
      FY(3)= GM(1)*TY2+GM(2)*TY3
      RETURN

C  INTERNAL SUBROUTINE FOR PRINTING PARTICLE POSITIONS AT A SPECIFIED
C     TIME STEP

      SUBROUTINE PRINT(N,X,Y)
      DOUBLE PRECISION X(3,3),Y(3,3)
      REAL XR(3),YR(3)
      DO 5 I=1,3
      XR(I)=X(I,3)
    5 YR(I)=Y(I,3)
      WRITE(6,99)N,(XR(I),YR(I),I=1,3)
   99 FORMAT(1X,I6,3X,3(2F13.L,3X))
      RETURN
      END
```

APPENDIX E

Fortran Program for General n-Body Interaction

```
C  INDEX TO PROGRAM VARIABLES

C    A = ALPHA
C    B = BETA
C    DT = TIME INCREMENT
C    EPS = CONVERGENCE CRITERION FOR NEWTON'S METHOD
C    FX(I) = FORCE COMPONENT ON PARTICLE I IN X-DIRECTION
C    FY(I) = FORCE COMPONENT ON PARTICLE I IN Y-DIRECTION
C    G = ATTRACTION PARAMETER
C    H = REPULSION PARAMETER
C    IEND = 0 IF ANOTHER DATA CASE FOLLOWS
C         = 1 IF END OF RUN
C    IMAX = MAXIMUM NUMBER OF ITERATIONS PER TIMESTEP FOR NEWTON'S
C           METHOD
C    INIT = USER-SUPPLIED SUBROUTINE TO CALCULATE INITIAL CONDITIONS
C    INP = 0 IF INITIAL DATA READ IN
C        = 1 IF CALCULATED IN SUBROUTINE
C    IPRINT = PRINT-STEP INCREMENT
C    IPUNCH = PUNCH-STEP INCREMENT
C    ISTART = 0 IF NEW DATA CASE
C           = 1 IF RESTART
C    JPUNCH = 0 IF NO PUNCH REQUIRED, PUNCH OTHERWISE
C    M(I) = MASS OF PARTICLE I
C    N = NUMBER OF PARTICLES IN SYSTEM
C    NFIX = TOTAL NUMBER OF PARTICLES TO BE FIXED
C    NMAX = MAXIMUM NUMBER OF TIMESTEPS THIS DATA CASE
C    NP(I) = NUMBER OF PARTICLE TO BE FIXED
C    NSTP = TIMESTEP NUMBER
C    OMEGA = SUCCESSIVE OVER-RELAXATION FACTOR FOR NEWTON'S METHOD
C    R(I,J,1) = DISTANCE BETWEEN PARTICLES I AND J AT PREVIOUS TIMESTEP
C    R(I,J,2) = DISTANCE BETWEEN PARTICLES I AND J AT CURRENT TIMESTEP
C    VX(I,1) = X-COMPONENT OF VELOCITY OF PARTICLE I, PREVIOUS TIMESTEP
C    VX(I,2) = SAME AS ABOVE, CURRENT TIMESTEP, PREVIOUS ITERATION
C    VX(I,3) = SAME AS ABOVE, CURRENT TIMESTEP, CURRENT ITERATION
C    VXO(I) = X-COMPONENT OF INITIAL VELOCITY, PARTICLE I
C    VY(I,1) = Y-COMPONENT OF VELOCITY OF PARTICLE I,
C    VY(I,2) =     SAME DEFINITIONS
C    VY(I,3) =        AS VX(I,J), ABOVE
```

```
C     VYO(I) = Y-COMPONENT OF INITIAL VELOCITY, PARTICLE I
C     X(I,1) = X-COMPONENT OF POSITION OF PARTICLE I,
C     X(I,2) =      SAME DEFINITIONS
C     X(I,3) =         AS VX(I,J)
C     XO(I) = X-COMPONENT OF INITIAL POSITION, PARTICLE I
C     Y(I,1) = Y-COMPONENT OF POSITION OF PARTICLE I,
C     Y(I,2) =      SAME DEFINITIONS
C     Y(I,3) =         AS VX(I,J)
C     YO(I) = Y-COMPONENT OF INITIAL POSITION, PARTICLE I

      IMPLICIT DOUBLE PRECISION(A-H,M,O-Z)
      DIMENSION XO(100),YO(100),VXO(100),VYO(100),X(100,3),Y(100,3),
     1VX(100,3),VY(100,3),FX(100),FY(100),NP(100),R(45,45,2),M(100)

 1001 FORMAT(8D10.0)
 1002 FORMAT(16I5)
 1003 FORMAT(4D10.0,I5)
 1004 FORMAT(I5,2D10.0)
 2000 FORMAT(1H1)
 2001 FORMAT(5X,'N',5X,'OMEGA',5X,'EPS',5X,'IMAX',/,I7,F10.4,E8.1,I9)
 2002 FORMAT(/6X,'A',8X,'B',8X,'G',8X,'H',7X,'DT'/4F9.3,E9.3)
 2003 FORMAT(' TIMESTEP ',6X,'M',9X,'X',14X,'Y'/2X,I6)
 2004 FORMAT(10X,F6.4,5F15.10)
 2005 FORMAT('  NON-CONVERGENCE AFTER ',I3,' ITERATIONS FOR TIMESTEP= ',
     1I6,' DT= ',E8.2)
 2501 FORMAT(2D25.18)

   10 PRINT 2000
      NSTP=0
      READ 1001,OMEGA,EPS,DT
      READ 1002,NMAX,IMAX,IPRINT,IPUNCH,JPUNCH,ISTART
      READ 1002,N,NFIX,INP
      READ 1003,A,B,G,H,IEND
      PRINT 2001,N,OMEGA,EPS,IMAX
      NM1=N-1
      IF(ISTART.EQ.0)GO TO 1
C  RESTART
      READ 1002,NSTP
      READ 2501,(XO(I),YO(I),I=1,N)
      READ 2501,(VXO(I),VYO(I),I=1,N)
      GO TO 3
C  NEW CASE
    1 IF(INP.NE.0)GO TO 2
      READ 1001,(XO(I),YO(I),I=1,N),(VXO(I),VYO(I),I=1,N)
      GO TO 3
    2 CALL INIT
    3 IF(NFIX.NE.0)READ 1002,(NP(I),I=1,NFIX)
      READ 1001,(M(I),I=1,N)
      PRINT 2002, A,B,G,H,DT
      OMW=1.0-OMEGA
      T=0.0
      DT2=DT/2.0
      PRINT 2003,NSTP
      DO 30 I=1,N
      PRINT 2004,M(I),XO(I),YO(I)
   30 CONTINUE
```

```
C  SPECIFY INITIAL GUESS FOR NEWTON'S ITERATION AT FIRST TIMESTEP
      DO 40 I=1,N
      X(I,3)=X0(I)
      VX(I,3)=VX0(I)
      Y(I,3)=Y0(I)
   40 VY(I,3)=VY0(I)
      CALL RCALC
C  UPDATE POSITIONS,VELOCITIES,DISTANCES FOR ALL TIMESTEPS
   45 NSTP=NSTP+1
      T=T+DT
      DO 60 I=1,N
      X(I,1)=X(I,3)
      VX(I,1)=VX(I,3)
      Y(I,1)=Y(I,3)
      VY(I,1)=VY(I,3)
      DO 50 J=1,N
   50 R(I,J,1)=R(I,J,2)
   60 CONTINUE

C  BEGIN ITERATION LOOP
      DO 90 K=1,IMAX
C  UPDATE ALL VARIABLES, CURRENT TIMESTEP, PREVIOUS ITERATION
      DO 70 I=1,N
      X(I,2)=X(I,3)
      VX(I,2)=VX(I,3)
      Y(I,2)=Y(I,3)
      VY(I,2)=VY(I,3)
   70 CONTINUE
C  UPDATE POSITIONS, CURRENT TIMESTEP, CURRENT ITERATION
      DO 73 I=1,N
      IF(NFIX.EQ.0)GO TO 72
      DO 71 J=1,NFIX
      IF(I.EQ.NP(J))GO TO 73
   71 CONTINUE
   72 X(I,3)=OMW*X(I,2)+OMEGA*(DT2*(VX(I,2)+VX(I,1))+X(I,1))
      Y(I,3)=OMW*Y(I,2)+OMEGA*(DT2*(VY(I,2)+VY(I,1))+Y(I,1))
   73 CONTINUE
      CALL RCALC
      CALL FCALC
C  UPDATE VELOCITIES, CURRENT TIMESTEP, CURRENT ITERATION
      DO 80 I=1,N
      IF(NFIX.EQ.0)GO TO 75
      DO 74 J=1,NFIX
      IF(I.EQ.NP(J))GO TO 80
   74 CONTINUE
   75 VX(I,3)=OMW*VX(I,2)+OMEGA*(DT*FX(I)+VX(I,1))
      VY(I,3)=OMW*VY(I,2)+OMEGA*(DT*FY(I)+VY(I,1))
   80 CONTINUE
C  TEST FOR CONVERGENCE
      DO 85 I=1,N
      IF(ABS(X(I,3)-X(I,2)).GT.EPS)GO TO 90
      IF(ABS(Y(I,3)-Y(I,2)).GT.EPS)GO TO 90
      IF(ABS(VX(I,3)-VX(I,2)).GT.EPS)GO TO 90
      IF(ABS(VY(I,3)-VY(I,2)).GT.EPS)GO TO 90
   85 CONTINUE
      GO TO 95
   90 CONTINUE
```

```
      PRINT 2005,K,NSTP
      GO TO 110
   95 IF(MOD(NSTP,IPRINT).NE.0)GO TO 103
      IF(JPUNCH.EQ.0)GO TO 102
      IF(MOD(NSTP,IPUNCH).NE.0)GO TO 102
      WRITE(1,1002)NSTP
      DO 151 I=1,N
      WRITE(1,2501)X(I,3),Y(I,3)
  151 CONTINUE
      DO 152 I=1,N
      WRITE(1,2501)VX(I,3),VY(I,3)
  152 CONTINUE
      IF(NFIX.NE.0)WRITE(1,1002)(NP(II),II=1,NFIX)
  102 CALL OUTP
  103 IF(NSTP.EQ.NMAX)GO TO 110
      GO TO 45
  110 IF(IEND.EQ.0)GO TO 10
      STOP

C  INTERNAL SUBROUTINE TO COMPUTE DISTANCES BETWEEN PARTICLES
      SUBROUTINE RCALC
      DO 210 I=1,NM1
      IP1=I+1
      DO 200 J=IP1,N
      R(I,J,2)=DSQRT((X(I,3)-X(J,3))**2+(Y(I,3)-Y(J,3))**2)
      R(J,I,2)=R(I,J,2)
  200 CONTINUE
  210 CONTINUE
      RETURN

C  INTERNAL SUBROUTINE TO COMPUTE FORCES (ACCELERATIONS)
      SUBROUTINE FCALC
      IA=A-1
      IB=B-1
      DO 600 I=1,N
      IF(NFIX.EQ.0)GO TO 450
      DO 400 K=1,NFIX
      IF(I.EQ.NP(K))GO TO 600
  400 CONTINUE
  450 SUMX=0.
      SUMY=0.
      DO 550 J=1,N
      IF(I.EQ.J)GO TO 550
      SUMG=0.
      SUMH=0.
      RIJ=R(I,J,1)+R(I,J,2)
      DO 500 I2=1,IA
      SUMG=SUMG+(R(I,J,1)**(I2-1))*(R(I,J,2)**(IA-I2))
  500 CONTINUE
      DO 501 I2=1,IB
      SUMH=SUMH+(R(I,J,1)**(I2-1))*(R(I,J,2)**(IB-I2))
  501 CONTINUE
      GD=(R(I,J,1)*R(I,J,2))**IA*RIJ
      SUMG=G*SUMG/GD
      HD=(R(I,J,1)*R(I,J,2))**IB*RIJ
      SUMH=H*SUMH/HD
      SUMX=SUMX+(SUMH-SUMG)*M(J)*(X(I,3)+X(I,1)-X(J,3)-X(J,1))
      SUMY=SUMY+(SUMH-SUMG)*M(J)*(Y(I,3)+Y(I,1)-Y(J,3)-Y(J,1))
```

```
  550 CONTINUE
      FX(I)=SUMX
      FY(I)=SUMY
  600 CONTINUE
      RETURN

C  INTERNAL PRINT SUBROUTINE
      SUBROUTINE OUTP
 3001 FORMAT(2X,I6)
 3002 FORMAT(15X,5F15.10)
      PRINT 3001,NSTP
      DO 800 II=1,N
      PRINT 3002,X(II,3),Y(II,3)
  800 CONTINUE
      RETURN

C  INTERNAL SUBROUTINE TO CALCULATE INITIAL CONDITIONS
C  FOR DISCRETE CONDUCTIVE HEAT TRANSFER
C    ANGLE(I) = ANGLE (IN DEGREES) OF INITIAL VELOCITY VECTOR WITH
C               RESPECT TO POSITIVE X-AXIS
C    IAXIS(I) = 0 IF LEFTMOST PARTICLE IN ROW I IS ON Y-AXIS
C             = +1,-1 IF LEFTMOST PARTICLE TO BE SHIFTED RIGHT OR LEFT,
C                     RESPECTIVELY
C    IROW(I) = NUMBER OF PARTICLES IN ROW I
C    IVEL(I) = NUMBER OF PARTICLE TO BE GIVEN AN INITIAL VELOCITY
C    NAXIS = NUMBER OF ROW ON X-AXIS
C    NROW = NUMBER OF ROWS IN SYSTEM
C    NVEL = TOTAL NUMBER OF PARTICLES TO BE GIVEN AN INITIAL VELOCITY
C    VEL(I) = MAGNITUDE OF INITIAL VELOCITY VECTOR

      DIMENSION IROW(5),IAXIS(5),IVEL(5),VEL(5),ANGLE(5)
 4001 FORMAT(16I5)
 4002 FORMAT(I5,2D10.0)

C  CALCULATE POSITIONS
      BASEX=(H*(B-1)/(G*(A-1)))**(1.0/(B-A))
      BASEY=SQRT(BASEX**2-(0.5*BASEX)**2)
      READ 4001,(IROW(I),IAXIS(I),I=1,NROW)
      IL=1
      IU=0
      DO 3 I=1,NROW
      IU=IU+IROW(I)
      XSHIFT=0.5*IAXIS(I)*BASEX
      JSHIFT=NAXIS-I
      DO 2 J=IL,IU
      X0(J)=(J-IL)*BASEX+XSHIFT
      Y0(J)=JSHIFT*BASEY
    2 CONTINUE
      IL=IU+1
    3 CONTINUE
C  CALCULATE VELOCITIES
   10 DO 4 I=1,N
      VX0(I)=0.0
      VY0(I)=0.0
    4 CONTINUE
      IF(NVEL.EQ.0)RETURN
      READ 4002,(IVEL(I),VEL(I),ANGLE(I),I=1,NVEL)
```

```
      PI=3.14159265358979324D+00
      RAD=PI/180.0
      DO 5 I=1,NVEL
      J=IVEL(I)
      THETA=ANGLE(I)*RAD
      VX0(J)=VEL(I)*COS(THETA)
      VY0(J)=VEL(I)*SIN(THETA)
    5 CONTINUE
      RETURN
      END
```

Answers to Selected Exercises

CHAPTER 1

1.5 It only prevents the circular process inherent in having to define further the words used in each definition.

1.6 $x - y$ is defined to be that number z which satisfies $x = y + z$.

1.7 $x \div y$ is defined to be that number z which satisfies $x = y \cdot z$.

1.8 No. The significance of the associative law is that it implies that it makes no difference as to which two numbers of the three are added first.

1.10 This axiom has caused so much controversy because inherent in it is the assumption that a straight line has an infinite length, while experimental scientists were constantly exposed to the finite structure of natural phenomena.

1.15 (a) Hint: List all possible arrangements and use elimination.

(b) Hint: Look at a checkerboard and observe the colors of the cut corners.

CHAPTER II

2.2 (a) $y_0 + y_1 + y_2 + y_3 + y_4 + y_5$

(b) $2y_1 + 2y_2 + 2y_3 + 2y_4 + 2y_5 + 2y_6$

(c) $y_{-1}^2 + y_0^2 + y_1^2 + y_2^2 + y_3^2 + y_4^2 + y_5^2$

(d) $y_2^2 - y_3^2 + y_4^2 - y_5^2 + y_6^2$

(e) $-2y_{-2} - y_{-1} + y_1 + 2y_2$

(g) $(-\sin y - 2 \cos x_{-1}) + (-2 \cos x_0) + (\sin y - 2 \cos x_1)$

$+ (2 \sin y - 2 \cos x_2) + (3 \sin y - 2 \cos x_3) + (4 \sin y - 2 \cos x_4)$

(h) $2\pi + 2\pi + 2\pi + 2\pi + 2\pi + 2\pi + 2\pi + 2\pi + 2\pi + 2\pi$

2.3 (a) $7\sum_{k=1}^{6} x_k - 9\sum_{k=1}^{6} y_k$

(b) $7\sum_{k=1}^{15} x_k^2 - 9\sum_{k=1}^{15} e^k$

(c) $5\sum_{k=0}^{20} \sin x_k$

(e) $\sum_{k=1}^{30} (-1)^k + 3\sum_{k=1}^{30} k^3$

(f) $3\sum_{k=-1}^{10} f_k(x) - 9\sum_{k=-1}^{10} g(k)$

(g) $\pi\sum_{k=0}^{10} \sin x_k + \sin \pi\sum_{k=0}^{10} x_k$

2.4 (a) $y_8 - y_0$

(b) $y_{11}^2 - y_{-1}^2$

(c) $y_{26}^3 - y_2^3$

(d) 11^2

(e) $\sin x_{11} - \sin x_1$

(f) 994^2

2.5 (a) 5

(b) 6

(c) 11

(d) $\sum_{k=1}^{5} 5 = 5\sum_{k=1}^{5} 1 = 25$

(f) $\sum_{k=0}^{10} \pi = \pi\sum_{k=0}^{10} 1 = 11\pi$

2.6 Hint: Use reasoning analogous to that for (2.21) and (2.23)

2.9 96 ft/sec

2.10 $K_k + V_k = 12800$, $k = 0,1,2,3,4.$

CHAPTER 3

3.1 (a) $x_0 = 0$, $x_1 = 0.25$, $x_2 = 0.50$, $x_3 = 0.75$, $x_4 = 1.00$, $x_5 = 1.25$

(b) $x_0 = 1$, $x_1 = 1.25$, $x_2 = 1.50$, $x_3 = 1.75$, $x_4 = 2.00$, $x_5 = 2.25$, $x_6 = 2.50$

(c) $x_0 = -1$, $x_1 = -0.75$, $x_2 = -0.50$, $x_3 = -0.25$, $x_4 = 0.00$, $x_5 = 0.25$, $x_6 = 0.50$, $x_7 = 0.75$

(d) $x_0 = -1$, $x_1 = -0.7$, $x_2 = -0.4$, $x_3 = -0.1$, $x_4 = 0.2$, $x_5 = 0.5$, $x_6 = 0.8$, $x_7 = 1.1$

3.2 (a) $x_k = \frac{k}{10}$, $k = 0,1,2,\ldots,10$

(b) $x_k = \frac{k}{5}$, $k = 0,1,2,\ldots,10$

(c) $x_k = \frac{7k}{10}$, $k = 0,1,2,\ldots,10$

(d) $x_k = -1 + \frac{2k}{5}$, $k = 0,1,2,\ldots,10$

(e) $x_k = \frac{k}{50}$, $k = 0,1,2,\ldots,100$

(f) $x_k = \frac{3k}{37}$, $k = 0,1,2,\ldots,37$

(g) $x_k = -3 + \frac{8k}{43}$, $k = 0,1,2,\ldots,43$

(h) $x_k = 0.2 + \frac{3k}{200}$, $k = 0,1,2,\ldots,100$

(j) $x_k = \frac{3k}{1000}$, $k = 0,1,2,\ldots,1000$

(k) $x_k = \frac{33k}{100000}$, $k = 0,1,2,\ldots,10000$.

3.7 (a) explicit

(b) explicit

(c) not a difference equation

(d) not a difference equation

(e) not a difference equation

(f) explicit

(g) not a difference equation

(h) not a difference equation

(i) implicit

(j) implicit

(k) implicit

3.16 Hint: Write out each side of the equation separately.

CHAPTER 4

4.3 For $k = 1$, $\cos\theta + i\sin\theta = \cos\theta + i\sin\theta$. Assuming the validity for $k = n$, we must show that

$$(\cos\theta + i\sin\theta)^{n+1} = \cos(n+1)\theta + i\sin(n+1)\theta.$$

Hence,

$$\begin{aligned}(\cos\theta + i\sin\theta)^{n+1} &= (\cos\theta + i\sin\theta)^n(\cos\theta + i\sin\theta)\\ &= (\cos n\theta + i\sin n\theta)(\cos\theta + i\sin\theta)\\ &= \cos n\theta\cos\theta - \sin n\theta\sin\theta\\ &\quad + i(\sin n\theta\cos\theta + \sin\theta\cos n\theta)\\ &= \cos(n+1)\theta + i\sin(n+1)\theta.\end{aligned}$$

4.4 (b) No motion.

(c) x_n is a cosine function with amplitude $\frac{\pi}{4}$, or 0.78540, and period 2π, or 6.28318.

4.6 $(\Delta t^2+4)v_{k+2} + 2(\Delta t^2-4)v_{k+1} + (\Delta t^2+4)v_k = 0$

4.7 (b) and (d) are periodic.

CHAPTER 5

5.2 (c)

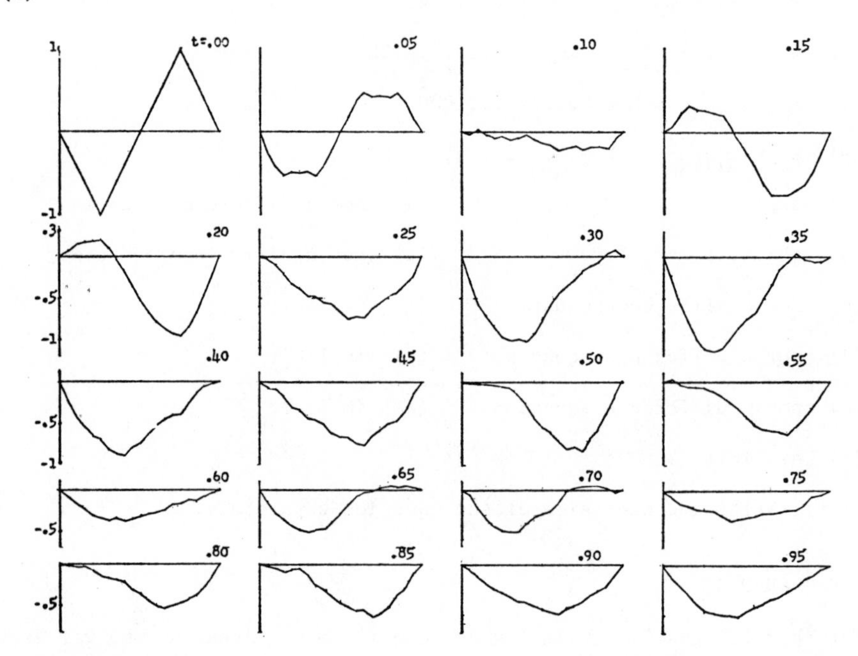

CHAPTER 6

6.1 (b) $r_1 = 13$, $r_2 = \sqrt{5}$, $r_3 = \sqrt{10}$

(c) (4, 14), (-2, 5), (5, 11), (-5, -13)

6.2 (b) $r_1 = 0$, $r_2 = 5$, $r_3 = 5$, $r_4 = \sqrt{2}$

(c) (-6, 8), (-1, 0), (-6, 8), (-6, 8), (-6, 9), (-12, 17), (-9, 14)

6.5 (b) $r_1 = \sqrt{29}$, $r_2 = \sqrt{29}$, $r_3 = \sqrt{29}$

(c) (0, 6, 8), (4, -6, 0), (2, 3, 12), (-2, -3, 4)

6.6 (b) $r_1 = 0$, $r_2 = \sqrt{2}$, $r_3 = \sqrt{29}$, $r_4 = \sqrt{29}$

(c) (2, 2, 0), (10, 15, 20), (19, 28, 36), (1, 1, 0), (5, 6, 4), (26, 42, 64

6.9 21, 21, 11, 3

6.10 0, 0, 5, 5

6.11 $\theta = 44°$ for $\vec{r}_1$ and $\vec{r}_2$

6.12 $\theta = 131°$ for $\vec{r}_2$ and $\vec{r}_4$; $\theta = 180°$ for $\vec{r}_3$ and $\vec{r}_4$

6.13 (a), (d), (e), (f)

6.16 (a) (-4, 8, -4) (d) (-4, 59, -33)

(b) (1, 0, -1) (e) (0, 0, 1)

(c) (-2, 4, -2) (f) (-8, 10, 12)

CHAPTER 7

7.4 (a)
$x_1 = 0.02,\ y_1 = 0$ $v_{1,x} = 0.2,\ v_{1,y} = 0$
$x_2 = 0.08,\ y_2 = 0$ $v_{2,x} = 0.4,\ v_{2,y} = 0$
$x_3 = 0.18,\ y_3 = 0$ $v_{3,x} = 0.6,\ v_{3,y} = 0$
$x_4 = 0.32,\ y_4 = 0$ $v_{4,x} = 0.8,\ v_{4,y} = 0$
$x_5 = 0.50,\ y_5 = 0$ $v_{5,x} = 1.0,\ v_{5,y} = 0$

(b) No motion.

(c)
$x_1 = 1.0000,\ y_1 = 1.0000,$ $v_{1,x} = 10.000,\ v_{1.y} = 10.000$
$x_2 = 3.1000,\ y_2 = 3.1000,$ $v_{2,x} = 11.000,\ v_{2,y} = 11.000$
$x_3 = 5.6100,\ y_3 = 5.6100,$ $v_{3,x} = 14.100,\ v_{3,y} = 14.100$
$x_4 = 8.9910,\ y_4 = 8.9910,$ $v_{4,x} = 19.710,\ v_{4,y} = 19.710$
$x_5 = 13.8321,\ y_5 = 13.8231,$ $v_{5,x} = 28.701,\ v_{5,y} = 28.701$

(d)
$x_1 = 0.0200,\ y_1 = 0.0200,$ $v_{1,x} = 0.2000,\ v_{1,y} = 0.2000$
$x_2 = 0.0804,\ y_2 = 0.0796,$ $v_{2,x} = 0.4040,\ v_{2,y} = 0.3960$
$x_3 = 0.1828,\ y_3 = 0.1772,$ $v_{3,x} = 0.6199,\ v_{3,y} = 0.5799$
$x_4 = 0.3303,\ y_4 = 0.3095,$ $v_{4,x} = 0.8554,\ v_{4,y} = 0.7434$
$x_5 = 0.5276,\ y_5 = 0.4716,$ $x_{5,x} = 1.1173,\ v_{5,y} = 0.8773$

7.5 (e)
$x_1 = .2000,\ y_1 = .2200,\ z_1 = -.2200$
$x_2 = .4008,\ y_2 = .4800,\ z_2 = -.4800$
$x_3 = .6056,\ y_3 = .7800,\ z_3 = -.7800$
$x_4 = .8211,\ y_4 = 1.1200,\ z_4 = -1.1200$
$x_5 = 1.0575,\ y_5 = 1.5000,\ z_5 = -1.5000$

$v_{1,x} = 1.0000,\ v_{1,y} = 1.2000,\ v_{1,z} = -1.2000$
$v_{2,x} = 1.0080,\ v_{2,y} = 1.4000,\ v_{2,z} = -1.4000$
$v_{3,x} = 1.0401,\ v_{3,y} = 1.6000,\ v_{3,z} = -1.6000$

$v_{4,x} = 1.1146, \quad v_{4,y} = 1.8000, \quad v_{4,z} = -1.8000$

$v_{5,x} = 1.2495, \quad v_{5,y} = 2.0000, \quad v_{5,z} = -2.0000$

7.6 (a) $K_0 = 0, \quad K_5 = \frac{1}{2}$

(b) $K_0 = 0, \quad K_5 = 0$

(c) $K_0 = 100, \quad K_5 = 823.747$

(d) $K_0 = 0, \quad K_5 = 1.0090$

7.7 (e) $K_0 = \frac{3}{2}, \quad K_5 = 4.781$

7.16 Let $r_k = r_{k+1} = r$. Then

$$\frac{Hm_1m_2[\sum_{j=0}^{m-2}(r_k^j r_{k+1}^{m-j-2})]}{r_k^{m-1} r_{k+1}^{m-1}} = \frac{Hm_1m_2(m-1)r^{m-2}}{r^{2m-2}} = \frac{Hm_1m_2(m-1)}{r^m} .$$

7.17 $r_{k+1}^m - r_k^m \equiv (r_{k+1}-r_k)(r_{k+1}^{m-1}+r_{k+1}^{m-2}r_k+r_{k+1}^{m-3}r_k^2+\ldots+r_k^{m-1})$

CHAPTER 8

8.3

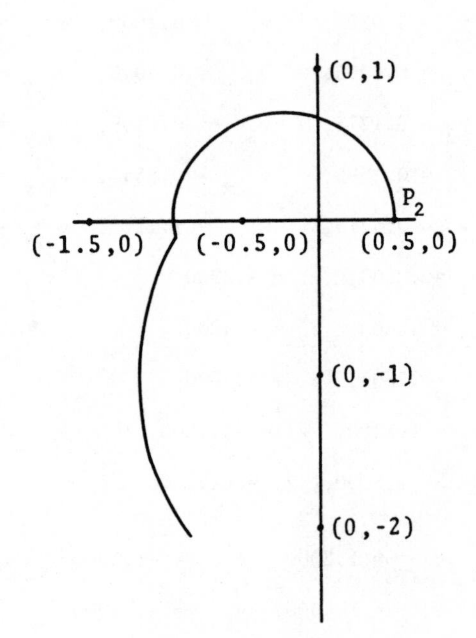

8.6 From each point hang also a heavy metal ball on a string to determine the vertical direction. Draw a vertical line across the figure at each hanging. The intersection of the two verticals is the center of gravity.

8.7 In Exercise 8.1 at t_0, $\bar{x} = 0$, $\bar{y} = 33\frac{1}{3}$.

8.9 For Exercise 8.1 (a), $m_1 v_{1,j,x} + m_2 v_{2,j,x} + m_3 v_{3,j,x} \equiv -10$,

$$m_1 v_{1,j,y} + m_2 v_{2,j,y} + m_3 v_{3,j,y} \equiv -100.$$

CHAPTER 9

9.1 (b) These are (8.12) - (8.14) and the corresponding formulas for the y-components.

9.2 2, $\sqrt{1.5}$, $(1.6)^{1/3}$, $(2.2)^{1/6}$, $(1.92)^{1/3}$

9.3 Hint: Let the potential energy V_k of the system at t_k be

$$V_k = \sum_{\substack{i,j=1 \\ i<j}}^{n} \left[\left(-\frac{G}{r_{ij,k}^{\alpha-1}} + \frac{H}{r_{ij,k}^{\beta-1}} \right) m_i m_j \right].$$

CHAPTER 10

10.7 Yes. The force of gravity results in an increase of particle compression as the distance from the top increases. The same is true for a gas.

10.9 The generation of heat is due to the transfer of large amounts of kinetic energy to module particles.

CHAPTER 11

11.1 It is almost impossible to keep the gyroscope's center of gravity over the string.

11.3 With spin, the sword will come down having the same direction as when it was thrown, making it easier to catch.

11.4 (a) $w_1 = w_2 = w_3 = w_4 = w_5 = 0$

(b) $w_1 = 1$, $w_2 = 2$, $w_3 = 3$, $w_4 = 4$, $w_5 = 5$

(c) $w_1 = -1$, $w_2 = -2$, $w_3 = -3$, $w_4 = -4$, $w_5 = -5$

11.5 Hint: Compare Theorem 3.2.

11.6 (a) $\alpha_0 = \alpha_1 = \alpha_2 = \alpha_3 = \alpha_4 = 0$

(b) $\alpha_0 = \alpha_1 = \alpha_2 = \alpha_3 = \alpha_4 = 1$

(c) $\alpha_0 = \alpha_1 = \alpha_2 = \alpha_3 = \alpha_4 = -1$

11.7 Hint: Compare Theorem 3.3.

11.8 (a) $v_1 = v_2 = v_3 = v_4 = v_5 = 0$

(b) $v_1 = 10$, $v_2 = 20$, $v_3 = 30$, $v_4 = 40$, $v_5 = 50$

(c) $v_1 = -10$, $v_2 = -20$, $v_3 = -30$, $v_4 = -40$, $v_5 = -50$

11.9 Hint: Your answer will depend on the radius.

11.10 Hint: Your answer will depend on the particular shape you choose.

CHAPTER 12

12.2 When the elevator is accelerating, the ball will not appear to fall. When the elevator is moving at a constant velocity, the ball will appear to fall under the influence of gravity.

12.3 (a) x' = 0, -1, -2, -3

(b) x' = 0, 1, 2, 3

(c) x' = 2, 5, 8, 11

(d) x' = -5, -11, -17, -23

12.4 (a) x = 0, 1, 2, 3

(b) x = 0, -1, -2, -3

(c) x = 1, -1, -3, -5

(d) x = -3, 1, 5, 9

12.5 (a) -1

(b) 1

(c) 3

(d) 1

12.6 (a) 1

(b) -1

(c) 5

(d) -10

References and Sources for Further Reading

Arons, A. B., and A. M. Bork (eds.) (1964). Science and Ideas. Prentice-Hall, Englewood Cliffs, N.J.

Auret, F. D., and J. A. Snyman (1978). Numerical study of linear and nonlinear string vibrations by means of physical discretization. Appl. Math. Modelling, 2, 7-17.

Born, M. (1951). The Restless Universe. Dover, New York.

Chebotarev, G. A. (1967). Analytical and Numerical Methods of Celestial Mechanics. Elsevier, New York.

Costabel, R. (1973). Leibniz and Dynamics. Cornell University Press, Ithaca, N.Y.

Courant, R., and H. Robbins (1941). What Is Mathematics? Oxford University Press, New York.

De Broglie, L. (1953). The Revolution in Physics. Noonday, New York.

Dettman, J. W. (1962). Mathematical Methods in Physics and Engineering. McGraw-Hill, New York.

Eddington, A. S. (1924). Space, Time and Gravitation. Cambridge University Press, New York.

Einstein, A., and L. Infeld (1938). The Evolution of Physics. Simon and Schuster, New York.

Feynman, R. D., R. B. Leighton, and M. Sands (1963). The Feynman Lectures on Physics. Addison-Wesley, Reading, Mass.

Forsythe, A. I., T. A. Keenan, E. I. Organick, and W. Stenberg (1969). Computer Science: A First Course. Wiley, New York.

Gottlieb, M. (1977). Application of computer simulation techniques to macromolecular theories. Comput. & Chem., 1, 155-160.

Greenspan, D. (1972). Numerical approximation of periodic solutions of van der Pol's equation. J. Math. Anal. & Appl., 39, 574-579.

Greenspan, D. (1973a). Discrete Models. Addison-Wesley, Reading, Mass.

Greenspan, D. (1973b). A finite difference proof that $E = mc^2$. Amer. Math. Mo., 80, 289-292.

Greenspan, D. (1974a). A physically consistent, discrete n-body model. Bull. Amer. Math. Soc., 80, 553-555.

Greenspan, D. (1974b). Discrete Newtonian gravitation and the three-body problem. Found. Phys., 4, 299-310.

Greenspan, D. (1980). Arithmetic Applied Mathematics. Pergamon, Oxford, England.

Heisenberg, W. (1958). Physics and Philosophy. Harper and Row, New York.

Hildebrand, F. B. (1952). Methods of Applied Mathematics. Prentice-Hall, Englewood Cliffs, New Jersey.

Hirschfelder, J. O., C. F. Curtis, and R. B. Bird (1954). Molecular Theory of Gases and Liquids. Wiley, New York.
Jammer, M. (1960). Concepts of Space. Harper, New York.
Jammer, M. (1962). Concepts of Force. Harper, New York.
Jammer, M. (1966). The Conceptual Development of Quantum Mechanics. McGraw-Hill, New York.
Kramer, E. E. (1970). The Nature and Growth of Modern Mathematics. Hawthorn, New York.
Krogdahl, W. S. (1952). The Astronomical Universe. Macmillan, New York.
Kuhn, T. S. (1962). The Structure of Scientific Revolutions. University Chicago Press, Chicago.
LaBudde, R. A., and D. Greenspan (1974). Discrete mechanics - A general treatment. J. Comp. Phys., 15, 134-167.
Larson, R. B. (1978). A finite particle scheme for three dimensional gas dynamics. J. Comp. Phys., 27, 397-409.
March, R. H. (1970). Physics for Poets. McGraw-Hill, New York.
Milne, W. E. (1949). Numerical Calculus. Princeton University Press, Princeton, N.J.
Newton, I. (1971). Mathematical Principles of Natural Philosophy (translated by A. Motte and revised by F. Cajori). University California Press, Berkeley, Calif.
Perry, J. (1957). Spinning Tops and Gyroscopic Motion. Dover, New York.
Potter, D. (1973). Computational Physics. Wiley, New York.
Rubinow, S. I. (1975). Introduction to Mathematical Biology. Wiley, New York.
Salacker, H., and E. P. Wigner (1958). Quantum limitations of the measurement of space-time distances. Phys. Rev., 109, 571-577.
Schubert, A. B., and D. Greenspan (1972). Numerical studies of discrete vibrating strings. TR 158, Dept. Comp. Sci., University of Wisconsin, Madison.
Stumpers, F. L. H. M. (1960). Balth. van der Pol's work on nonlinear circuits. IRE Trans. on Circuit Theory, Vol. CT-7, No. 4, 366-367.
Taub, A. H. (1971). Studies in Applied Mathematics. Math. Assoc. Amer., distributed by Prentice-Hall, Englewood Cliffs, N.J.
Taylor, E. F., and J. A. Wheeler (1966). Spacetime Physics. Freeman, San Francisco.
von Karman, T. (1963). Aerodynamics. McGraw-Hill, New York.
Whitrow, G. J. (1961). The Natural Philosophy of Time. Harper and Row, New York.
Young, J. W. (1911). Lectures on Fundamental Concepts of Algebra and Geometry. Macmillan, New York.

Index